Technologies for Plasma Renovation of the Phase Structure of Substandard Turbine Blades of CHP

Tatyana Yurievna RATUSHNAYA
Vitaliy Vladimirovich SAVINKIN
Andrei Victor SANDU
Petrica VIZUREANU

Published by **Materials Research Forum LLC**
Millersville, PA 17551, USA

Published as part of the book series
Materials Research Foundations
Volume 150 (2023)
ISSN 2471-8890 (Print)
ISSN 2471-8904 (Online)

Print ISBN 978-1-64490-264-6
ePDF ISBN 978-1-64490-265-3

Distributed worldwide by

Materials Research Forum LLC
105 Springdale Lane
Millersville, PA 17551
USA
http://www.mrforum.com

Printed in the United States of America
10 9 8 7 6 5 4 3 2 1

Table of Contents

Designations and Abbreviations

CG - carrier gas flow

CHP – combined heat and power plant

CIS – Commonwealth of Independent States

EMF - Electromotive force

ECM - eddy current methods

GAP - gas-air path

MMM - magnetic memory method

MPD - magnetic particle flaw detection

RD - regulatory documents

VC - visual control

UT - ultrasonic testing

Introduction

Currently, despite the active use of alternative energy sources, strategic facilities such as CHP are of particular interest to world powers in the field of energy and heavy engineering. Increasing their energy efficiency, productivity and ensuring trouble-free operation today continues to form the basis of industrial and technological development. However, the trends in price policy changes for energy resources, repairs and maintenance are subject to the influence of global monopolists. The problem of dependence on other countries can only be solved by developing an import substitution strategy through the development of recovery production by developing an energy-efficient plasma technology for the renovation of substandard (currently discarded) turbine blades of CHP.

The book contains a comprehensive and interdisciplinary study in the field of plasma recovery of substandard CHP turbine blades. Therefore, the advantages and disadvantages of recovery methods will be presented, as well as the traditional methods that are used in production, as well as the latest achievements of scientists in the field under consideration.

The developed algorithm and technology for the restoration of substandard CHP turbine blades by the method of implantation allows expanding the range of parts to be restored and ensuring the restoration of the structure of highly loaded parts, which will significantly increase the efficiency of CHP turbines, while reducing the cost of electricity generated. The proposed algorithm for choosing a technology for the restoration of substandard CHP turbine blades, as well as reasonable indicators of the quality of the technological process due to the modification of structural-phase transformations in the material of martensitic and austenitic classes, make it possible to control the physical and mechanical properties of the coating.

Thus, to improve productivity, reduce production costs, increase the durability of energy-loaded equipment will contribute to an efficient production process.

The book is relevant for fundamental and applied research in the field of engineering and energy, as it presents cutting-edge knowledge in the stated areas. Therefore, the book titled "Technologies for Plasma Renovation of the Phase Structure of Substandard Turbine Blades of CHP" is a milestone, practical and theoretical guide for both specialists in this field and young researchers who want to deepen their knowledge in the field of plasma spraying.

The market for restoration and diagnostic services is poor in offering a qualitative assessment of the restored parts of power plants due to stringent requirements. Kazakh thermal power plants are strategic facilities, and foreign firms do not always have the opportunity to participate in the competition for diagnosing and forecasting, and if they meet the requirements, they offer an inflated price for their services, which ultimately increases the cost of electricity generated.

A wide range of proposed recovery methods and technologies is usually based on thermodynamic action with the introduction of fluxes and additives into the weld pool. It is known that any thermal operation inevitably entails the formation of internal stresses of the metal, the magnitude of which depends on the coefficient of thermal expansion.

These factors affect the modification of the physical and mechanical properties of the formed coating, and lead to degradation of the original phase structure. The main ways to improve their quality are described by leading scientists Klubnikina V.S. [1], Kostikova V.I., Shesterina Yu.A. [2], Lashchenko G.I. [3] and others who have made a significant contribution to the development of welding technologies. However, existing technologies are aimed at restoring the geometric parameters of parts and the mechanical properties of their surface. From the works [4-8], it was found that the base of the part perceives the dynamic load of cyclic action, and its surface reacts to the action of an aggressive environment.

To date, despite advances in the field of recovery technologies, the direction on the method of forming the optimal phase structure of turbine blades from steels of austenitic and martensitic classes has not been sufficiently studied. Consequently, there is a scientific problem in the formation of a knowledge base and the establishment of dependences of the influence of the concentration of internal stresses on the phase composition of the structure with various variations in the technological modes of restoration.

Chapter 1. Analysis of the Performance of the CHP Turbines and the Prospects for Increasing Their Life

1.1   Analysis of Operating Conditions and Design and Technological Features of the Turbine Blades of the CHP

The main type of engine in modern CHPs is a steam turbine. The rotor blades, being the working body of the turbine, are subjected to dynamic loads during operation, which causes their wear. Depending on the type of defect and the severity of its consequences for the structure as a whole, there are various ways to restore turbine blades. However, these methods do not always allow to achieve the necessary physical and mechanical properties that a turbine blade should have. Therefore, a scientific and technical problem arises, which is to expand the range of defective blades suitable for the restoration procedure.

For the production of thermal and electrical energy in the energy sector, turbines are used with rotational movement of the working body of the rotor with blades, driven by a steam pressure flow formed as a result of the combustion of crushed coal, or by a water flow. Thus, there is a transformation of one type of energy (pressure of steam, water) into another (heat and electricity) [9 p. 345; 10].

The steam turbine is the main type of engine in modern thermal and nuclear power plants. The advantage of using a turbine is that it operates at high speed, produces a large amount of power (more than 1000 MW) with a relatively small size and weight. At the same time, the steam turbine has high technical and economic indicators: a relatively low unit cost, efficiency, reliability, and a service life of tens of years [11–13].

The steam turbine is classified as a blade engine [14-16]. Its working body is a disk mounted on a shaft with a crown of curved blades. A number of simple or combined nozzles are located in front of the blades. The nozzles are the fixed part of the turbine; they are attached to the body or diaphragm (Figure 1)

The principle of operation of the turbine is characterized by two main processes (adiabatic expansion of gas and conversion of kinetic energy into mechanical and electrical energy), which occur in the nozzle arrays and channels formed by the working blades, when the working fluid passes through them - steam or gas [17-18].

When burning various types of fuel in the furnace, water turns into steam. With further overheating of the steam to 435 °C and a pressure of 3.43 MPa, the steam is transferred through pipes to the turbine, where it is uniformly distributed over the nozzles with the help of special parts. Thus, in the nozzles, the potential energy of the steam is converted into kinetic energy; in the blade channel under the influence of a jet of steam, a centrifugal force arises that acts on the blades and causes rotation of the turbine rotor. When the shaft is connected to a current generator, mechanical energy is converted into electrical energy.

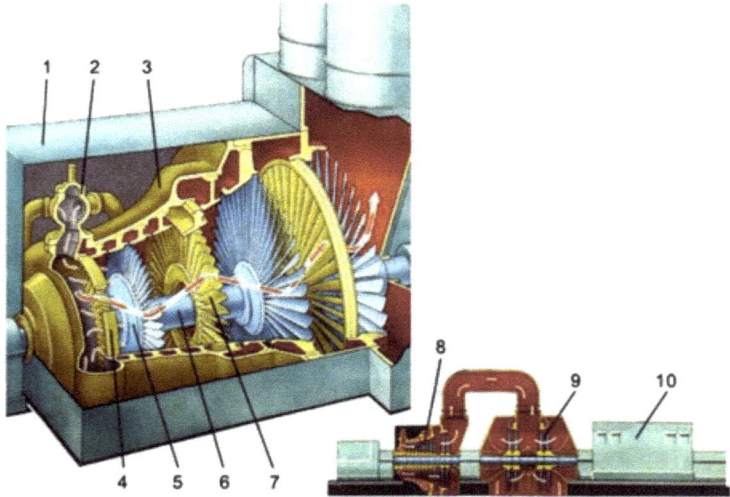

1 - casing, 2 - steam distribution device, 3 - turbine housing, 4 - nozzle apparatus, 5 - rotor disk, 6 - turbine shaft, 7 - stator (guide vane) disk, 8 - high pressure cylinder, 9 - low pressure cylinder, 10 - generator

Figure 1. Scheme of the steam turbine

The main working body of the turbine is the blades. Their parameters (profile shape, geometric dimensions, materials used) depend on the operating conditions of the blades in a multistage turbine and are very diverse [19–22]. The rotor blades have a complex curvilinear profile and various geometric dimensions, depending on the turbine stage in which they are used. The asymmetry of the blade profile forms curvilinear tapering channels. When leaving the nozzle, the steam in the guide channels of the tapering part of the blade continues to expand. This effect provides an increase in relative speed. This causes a reactive pressure force acting on the rotor blades.

The design of rotor blades can be conditionally represented as consisting of three main parts [23]: the tail, the working part of the airfoil, which has the leading and trailing edges of the head. Each of these parts has a large number of design variations (Figure 2).

The choice of the shank type is determined by the loads acting on the blade, which are perceived by the shank and transferred to the disk [24-26].

The highest load-bearing capacity has fir-tree type liners with end winding, used for the longest and most loaded blades of the last stages.

These liners are often made (in plan) along an arc of a circle, which makes it possible to avoid hanging the edges of the root section profile beyond the limits of the shank and facilitates the winding of the blades into the disk (Figure 2).

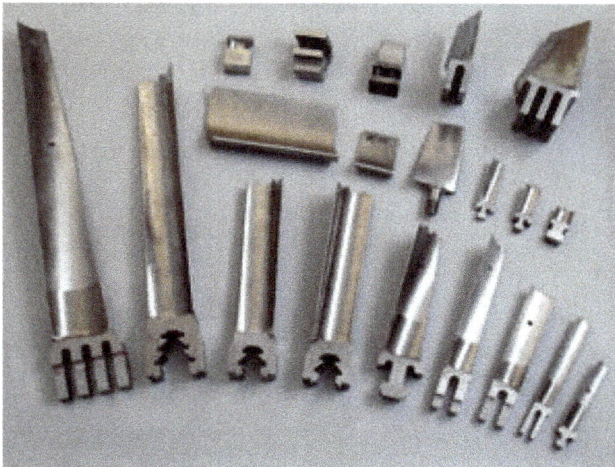

Figure 2. Steam and gas turbine blades

According to the results of the research, it was found that the main power unit of steam and gas turbines is a rotor with a set of working blades of complex design geometry. It has been established that the main technological and operating parameters largely depend on the operating conditions of the turbine and the technical condition of the blade apparatus. The reliability of the blade apparatus is ensured and regulated by a system of diagnostics, maintenance, and repair and restoration processes. Therefore, a theoretical assumption can be made that the reliability of the turbine will depend on the quality indicators of the turbine blades, which are provided by a complex technological process of repair and restoration of worn surfaces of a curved profile.

Thus, in the course of research, it is necessary to prove that the use of blades restored by a plasma energy source will reduce the inertia of steam plants, reduce the cost of generated electrical energy without losing energy quality, and also reduce the cost of repairing steam turbines.

1.2   Types of Failures and Consequences of Accidental Defects in Turbine Blades of CHP

The analysis of CHP turbine failures was carried out on the basis of information provided by the thermal power plant of the city of Petropavlovsk CHP-2 - failure investigation reports and other reporting forms (Figure 3) [27].

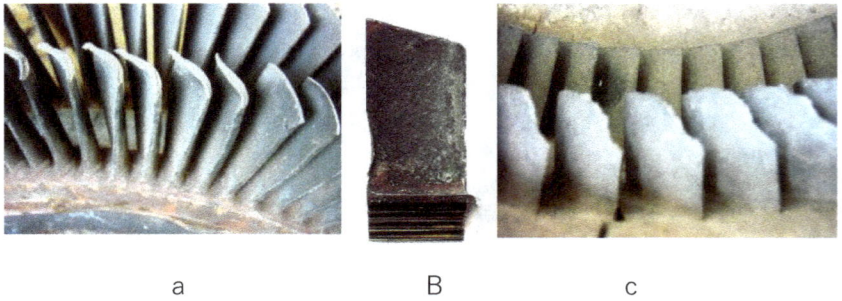

a   B   c

a - deformation of the rotor blades, b - turbine blade after operation,
c - deformation of the stator blades

Figure 3. Deformation of the blades during operation

It has been established that at many CHPs, the results of monitoring turbine failures are not filled in due to the laboriousness of the process and not all actual damage to the nodes is indicated. The method for fault detection of steam turbines at CHP plants is regulated by RD 153-34.1-17.462-00 "Methodological guidelines on the procedure for assessing the performance of steam turbine blades during manufacture, operation and repair", but in practice, for a number of subjective reasons, it is not always observed.

Despite this, the results of statistical analysis of data received from power plants provide sufficient information to identify the most common defects. Using a deterministic method for assessing defects, the reliability of the results of such an analysis largely depends on the number of objects for which information was collected.

The book summarizes the results of a statistical analysis of the causes of turbine failures from among those subjected to restoration work that caused the shutdown of the turbine unit.

In the study of the causes of failures, the results were used, both with factory blades and with remanufactured ones. The failure rate of turbines of various types is presented in Table 1. At present, the issue of increasing the vibrational reliability of a turbine with the use of remanufactured blades has not been sufficiently studied. Thus, it is necessary to tighten the requirements for the development of technology for the restoration of physical and mechanical properties, providing a resource of the restored blade, close to the new one.

Table 1. Frequency analysis of shutdowns based on turbine type

Turbine type	Number of stops	Turbine type	Number of stops
K-800-240 LMZ	19	K-300-240 HTZ	76
K-500-240 LMZ	6	K- 150- 130 HTZ	84
K-300-240 LMZ	89	T-250/300-240 TMZ	29
K- 100-90 LMZ	95	T-175-130 TMZ	15
T- 180/210- 130 LMZ	17	T–100/120	190
K-500-240 HTZ	10	ПT-135-130 TMZ	31

More than 70% of shutdowns are caused by various damages to the flow path of the turbine [28, 29], which occur due to increased vibrations, we will analyze them in more detail. These damages are associated primarily with the wear of various parts of the turbine rotor, the stratification for which is given in Table 2. Analyzing damage to the rotor, its constituent elements were considered: the rotor shaft, couplings, disks, seals, rotor blades. The number of shutdowns due to the breakdown of the above parts, as well as their share ratio, are shown in the table.

Table 2. Rotor stall causes stratification

Reasons for shutdowns	Working blades	Rotor shaft	Coupling	End compactions	Diaphragm and shroud seals	Shroud	Disk
Number of stops	110	24	18	20	14	36	14
Share in the total set %	46,6	10,2	7,6	8,5	5,9	15,3	5,9

Using the data obtained, a Pareto analysis was carried out. Based on the results of which, a distribution diagram of the causes of turbine shutdowns was constructed (Figure 4) [30-32].

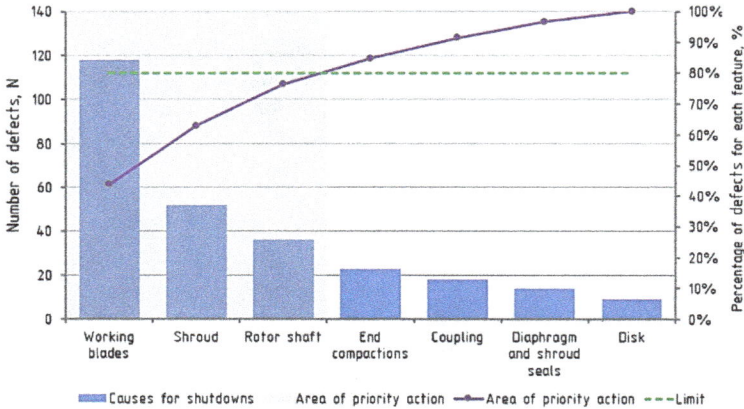

Figure 4. Pareto Chart [27 p. 277]

As can be seen from the diagram, the most frequent turbine failures occur due to the failure of their working blades, shroud and end seals. In order to determine the possible causes of breakdowns, we will perform a more detailed analysis of the first three groups.

Based on the results of scheduled inspections, an analysis was made of the causes of failures of the restored rotor blades.

According to the data obtained, the main share of damages is the failures of rotor blades (46.6%). The percentage of failures of the rotor shaft is 10.2%, it should be noted that 75% of the damage to the rotor shaft is caused by shaft deflection, which subsequently leads to rubbing in the flow path and damage to the ridges of the end seals.

Next, consider the assembly unit with blades - "the last stage of the turbine". The main causes of damage and, as a result, turbine shutdown are shown in Figure 5a.

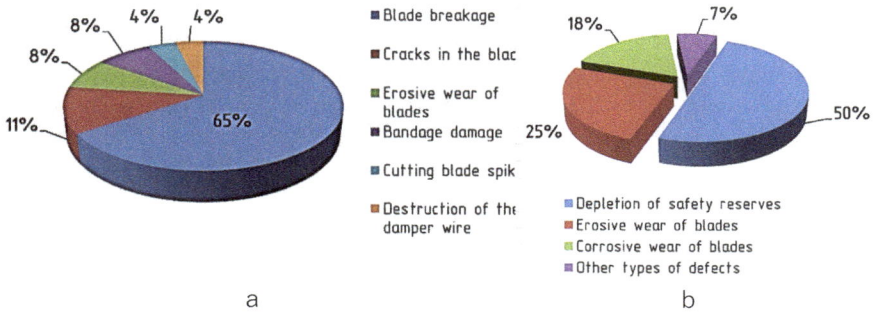

a - the main causes of damage to the scapular apparatus;
b - distribution of types of wear of rotor blades

Figure 5. Causes of defects

According to the analyzed data, the main part of the damage to the working blades (up to 65%), Figure 5 a, is caused by the breakage of the root of the working blade at the root section and in the cross-sectional area of the holes intended for the damper wire.

Also, according to the results of the study of the data obtained, cracks were found in the blade (about 12% of the total number of all identified damage to the blades) and significant corrosion-erosion wear, which amounted to approximately 8% of the total number of damages to the rotor blades.

During the analysis of the causes of blade defects, it was found that most of them are associated with the exhaustion of the safety margin and metal fatigue (50% of damage). The next important cause of defects is the erosive wear of the edges of the blade airfoil (25% damage).

After studying the causes of defects, the following diagram was obtained (Figure 5b).

1.3 Modern Turbine Blade Recovery Technologies

Methods for restoring the structure of CHP turbine blades, used in world practice, have not yet been widely used.

According to research [33 p. 37, 34-37] substandard blades are replaced with new ones, which leads to significant material costs. Replacing worn blades with new ones is associated with high costs, long delivery times and significant labor costs. At present, there is no analogue of the technology for the restoration of substandard blades in Kazakhstan.

As a rule, the last rows of low-pressure rotor blades are cut off, which leads to a decrease in efficiency and an increase in the cost of electricity [35 p. 29].

The requirements for remanufactured steam turbine blades at CHPs include their compliance with regulatory and technical documentation (overall dimensions, longitudinal and cross-sectional profile, physical and mechanical properties, such as hardness, fatigue resistance, strength characteristics, corrosion and erosion resistance).

Currently, various methods are used to restore substandard turbine blades: galvanization, spraying, grinding, welding, surfacing, deformation, etc.

There is a traditional method for repairing the working blade of a steam turbine [38]. When repairing a blade using this method, a plate is welded to the place of the worn edge section. The main disadvantage of the proposed repair method is the complexity of manufacturing and welding of the plate, as well as the high level of tensile residual stresses that arise after welding of the plate and welding of protective overlays.

In the studies of Godovskaya G.V., Isanberdin A.N., Lyudvinitsky S.V., Khafizov R.Kh. A method has been developed for restoring turbine blades, including the removal of damaged material, surfacing of the repaired area, heat treatment to relieve residual stresses, and mechanical processing of the blade [39].

The disadvantage of this method is the use of dissimilar materials for surfacing, which, on the one hand, provides a combination of ductility of the deposited material and high hardness of its surface, but at the same time leads to the formation of a heterophase layered structure, which reduces corrosion resistance.

The method for restoring the working blades of steam turbines proposed by Lappa V.A., Fedin I.V., Khromchenko F.A. stellite plates [40].

The disadvantages are the need to remove the blades from the rotor, the need for heat treatment after surfacing, which complicates the work and increases its cost. A significant disadvantage of this method is the lack of heat treatment after welding of protective overlays, as a result of which a high heterogeneity of the structural-phase composition and a high level of tensile residual stresses remain in the blade material. Researcher Gonserovsky F.G. [41 p. 41; 42, 43] during surfacing and welding in the heat-affected zone, it was also not possible to reduce tensile - residual stresses and high heterogeneity of the structural-phase composition of the material of the restored blade. Tensile residual stresses reduce the fatigue resistance of the material, and the structural-phase inhomogeneity of the material reduces its corrosion resistance, since the combination of sections with different electrode potentials forms many microgalvanic couples.

All of the above methods are aimed at restoring the geometric shapes and physical and mechanical properties of the working surface. The difference between them lies in the technological modes, the operating material used, and the equipment or complexes used.

According to the results of the analysis of known achievements in the field of restoration processes of steam turbine blades, it has been established that all technological achievements are aimed at restoring the geometric parameters of the blades and modifying the contact surface of the leading and trailing edges of the blade airfoil.

However, the technical and scientific problem of repairing substandard blades is much broader. The specificity of the established classification of wear types and associated defects showed that the durability of rotor blades is largely determined by the structural-phase components and interatomic bonds of the material. Therefore, the restoration of the physical and mechanical properties of the blade microstructure is a priority direction for renovations in the energy and mechanical engineering industries. This feature is due to the fact that only some properties (wear resistance, corrosion resistance, heat resistance, etc.) are imparted to the surface layer of the restored part; the basis of the part perceives the dynamic and vibration load itself.

Therefore, a hypothesis has been put forward that the provision of strong interatomic bonds in various phases of the microstructure of the part will increase the durability of the blades under cyclically changing dynamic and vibration loads. In practice, a change in the phase structure occurs when exposed to high temperatures. This problem can be solved by technological methods using highly concentrated sources of plasma energy of local impact.

Thus, the restoration of CHP turbine blades using a source of laser-plasma energy is an urgent task.

Currently, in the CIS countries, plasma and laser technologies for the restoration of power equipment using plasmatrons, multi-composite powders and wires are most widely used [38 p. 57].

Research and development to improve the mechanical properties of products made of nickel alloys, carried out under the leadership of Nagirny L.A. Prikhodko T.V. [39 p. 2, 43 p. 11, 44], are not effective for loaded turbine blades. The main disadvantage is the need to control the microstructure of the blade base material and maintain a balance of tensile and compressive stresses with a complex design geometry of the turbine blades. In this regard, at present, restoration technologies lag far behind modern realities [38 p. 65].

The most relevant method is prosthetics using highly concentrated sources of plasma energy. Research Frenkel Ya.I., Dovbysh V.N. and others [44 p. 012037-2, 45, 46] have shown that plasma technologies make it possible to control the graininess of the microstructure and control the martensitic and austenitic environment of the material, and not only limit the stresses arising in the blades, but also use their effect (compression - tension).

We singled out three main types of welding of parts of complex geometry: surfacing under a layer of flux; vibro-arc surfacing; plasma surfacing-spraying.

But all of them have a number of significant drawbacks: complex equipment, scarce raw materials and low quality of surfacing. A common and significant drawback of all recovery methods is the significant heating of parts, poor wear resistance, as a result of which fatigue strength decreases and their design geometry is violated [41 p. 41; 40 p. 245; 42 p. 67; 43 p. 5; 44 p. 012037-3; 45 p. 47; 46 p. 33]. Fatigue defects are caused by the presence of tensile stresses that arise during heating of individual sections of the metal and structural stresses, the formation of which is associated with an increase in volume during the transition of austenite to martensite. An analysis of the technical and economic indicators of existing methods for restoring blades is presented in Table 3 [47]

Table 3. Feasibility study of methods for restoring turbine blades

Technology characteristic	Welding	Surfacing	Spraying	Finishing - plasma hardening	Quench hardening
Processing scheme					
Thickness of processed blades, mm	0.5 - 10	over 2	any	any	over 3
Coating thickness (or hardening depth without flashing), mm	-	large (1-4)	medium (0.1-1.0)	small (0.0005-0.003)	medium (0.3-1.5)
The strength of the connection between the coating and the base	-	high	reduced	high	–
Integral base temperature, °C	high (200-1000)	high (200-1000)	low (100-200)	low (100-200)	low (200-300)
Thermal deformation of the product	reduced	yes	no	no	yes
Structural changes in the base	yes	significant	no	minimal	yes
Preliminary preparation of the base surface	descaling and organic cleaning	descaling and organic cleaning	abrasive blasting	cleaning from organics (degreasing)	descaling and organic cleaning
Coating porosity	-	no	yes	мини-мальная	–
Preservation of the surface roughness class	-	no	no	yes	yes
The surface may have increased hardness	-	yes	yes	yes	yes
Coating can be wear resistant	-	yes	yes	yes	–
The coating can be heat resistant (up to 1000 °C)	-	yes	yes	yes	–
Coating can be dielectric	-	no	yes	yes	–
Ability to maintain high hardness of the base	no	limited	yes	yes	yes (outside the heat affected zone)
The possibility of carrying out the technical process manually and automatically	mostly automatic	yes	yes	yes	only automatically

From Table 3 it can be seen that the current technology for the restoration of blades is plasma spraying of the working surface. The recommended recovery method provides more efficient sintering of metal particles with a restored surface and the necessary wear resistance, and by correcting the regime characteristics of the recovery process, it is possible to significantly improve the physical and mechanical properties of the part.

Of the variety of methods for restoring blades, automatic surfacing under a flux layer [48] and plasma spraying [49] have received wide practical application. This fact can be explained by a number of advantages compared to other recovery methods. These advantages include: the possibility of obtaining a layer of optimal chemical composition and properties; the possibility of coating flat surfaces, bodies of revolution and curved surfaces; the resulting coatings do not require subsequent heat treatment; the average porosity of plasma coatings varies from 3 to 20%; the possibility of automating the surfacing process; wide limits of variation of technological parameters of recovery; high efficiency of the plasma jet.

Plasma coatings have higher physical and mechanical properties compared to coatings deposited by other metallization methods [38 p. 132; 46 p. 47]. During the process of melting the coating, a liquid phase is formed, which provides a more intensive course of diffusion processes. Thereby increasing adhesion, increasing the mechanical strength and wear resistance of the coating. The wear resistance of the melted coating is 1.5-2 times higher than the wear resistance of steel 45, hardened to a hardness of 45 to 55 HRC.

For plasma technology, in contrast to argon-arc surfacing [47 p. 135; 50, 51]. are characterized by a minimum allowance for subsequent machining (about 200 µm), a narrow heat-affected zone (up to 100 µm), the presence of a fine-grained structure of the deposited layer, minimal (local) energy input, an increase in the repair area of the blade surface, the absence of heat treatment, increased mechanical characteristics of the deposited layer.

It is also necessary to note the flexibility of the process, which makes it possible to use both metal powder and wire as a filler material [52–54].

The solution of the scientific and practical problem of improving the reliability of the turbine blades of CHPs requires the justification of priority design parameters that affect the reduction of dynamic loads of the blades. To increase the efficiency of the recovery technology, the proposed range of design parameters of turbine blades must be integrated with the parameters of the technological process for the restoration of worn turbine blades. One of the mandatory stages in the implementation of this task is the establishment of a causal relationship of the influence of technological parameters on the reduction of dynamic loads in the restored turbine blades (Figure 6).

Leading scientists of Russia and Kazakhstan [37 p. 20; 39 p. 1; 40 p. 325; 42 p. 5; 43 p. 7; 55, 56] set and solved the problems of increasing the durability and performance of parts operating with alternating loads.

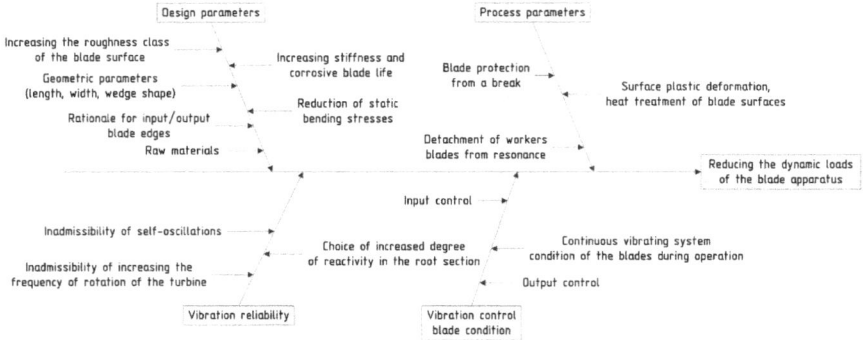

Figure 6. Cause-and-effect diagram of the influence of technological parameters on the reduction of dynamic loads in water

Currently, researchers pay insufficient attention to establishing patterns, the influence of various factors on wear resistance, adhesion strength to the base metal, fatigue strength, etc. Adhesion strength and hardness of coatings formed by plasma spraying is the weakest point in the characteristics of this method of restoring worn parts [38 p. 115]. These factors do not allow the most efficient use of the entire hidden potential of recovery by promising plasma technologies. From analytical studies it has been established that among the wide variety of proposed recovery methods there are no clear rules, criteria and methods for substantiating and choosing a recovery method. Technology selection algorithm proposed by Savinkin V.V. and others in [47 p. 205] solves the technical problem of providing manufacturing enterprises with a unified methodology for justifying and choosing a restoration method, taking into account the specifics of restoration production. The development of this recovery technology selection algorithm was carried out taking into account resource saving.

Authors [38 p. 125] propose to analyze technologies in two stages. The first step in recovery technology is to evaluate the manufacturability of the process in accordance with quality indicators. Input data are substandard CHP turbine blades and technologies that are planned to be applied for recovery.

Also, in the course of the analysis, it is revealed which technologies will make it possible to qualitatively restore performance [38, p. 68].

At the second stage, a quantitative analysis of manufacturability is carried out.

The algorithm for choosing the optimal restoration technology, where the cost price and ensuring the properties of parts, taking into account defects in their surfaces, is based on the principle of performing restoration at the lowest cost in cases where it is possible.

The developed algorithm for the restoration of parts allowed the authors to increase the efficiency of restoration production, since, along with an economic assessment, it is used to select the optimal technology not only in terms of economy, but also, which is most important in restoration, ensuring the working properties of parts. However, this algorithm does not take into account the possibility of restoring substandard turbine blades, which are usually rejected. Taking into account the results of the analysis, it can be assumed that the range of blades suitable for restoration can be expanded by introducing criteria for assessing the mechanical properties of the blade, which inevitably requires an integrated approach and improvement of the blade fault detection technique.

Taking into account the technological features of plasma reduction production technologies, it is a difficult technical task to synthesize the advantages of laser and plasma reduction technologies based on the use of highly concentrated energy sources. When developing innovative restoration technologies, it is necessary to clearly and unambiguously form the criteria and indicators of the quality of restoration of blades of complex geometry.

The idea of developing an innovative technology is to restore a working blade that has defects in which it was previously rejected. In this case, the method of prosthetics using plasma technology will be applied. Such a synthesis of technological processes is considered for the first time in Kazakhstan and has not previously been used in the practice of the CIS countries.

1.4   Analysis of Methods for Assessing the Quality of Restoration of Complex Geometry Blades

According to the results of various studies [57-59], it was established that during the operation of CHP turbines, zones of stress concentration are created in the working blades, which reduce the fatigue strength of the restored sections of the blades operating under conditions of alternating cyclic loads. For rotor blades restored by plasma surfacing and spraying, it is mandatory to conduct studies on the concentration of internal stresses in the material structure.

The control of the restored parts includes: external inspection, measurement of the coating thickness, determination of porosity, inclusions, discontinuities, hidden cracks (metallographic studies), adhesion of the coating (by the "pin" method), and erosion and corrosion resistance.

According to [60], the control process can be carried out by the following main methods:

- visual control (VC);
- magnetic particle flaw detection (MPD);
- ultrasonic testing (UT);
- endoscopy without opening and with opening of the cylinder;
- vibration control on a working turbine using a discrete-phase method;
- vibration tests on the rotor extracted from the turbine;
- acoustic emission control;
- diagnosing with the help of samples-witnesses of fatigue failure.

The most widely used in production practice, despite the known shortcomings, received ultrasonic testing.

The main means of control recognized by the world community of the curved profile surfaces of the blades are control and measuring machines. To control the seats of the blade shank with a complex relief, special measuring instruments have been developed.

A group of non-destructive testing methods, such as color flaw detection, luminescent method, acoustic emission method, make it possible to detect microdefects at a shallow depth of the material under study.

The set of microroughnesses of the machined surfaces of the blades is checked by a known method using samples or universal measuring instruments.

When restoring substandard CHP turbine blades, it is necessary to carry out input and output quality control. The methods regulated by [60 - 63] do not give a complete picture of the state of the blade metal. This problem can be solved using modern methods of flaw detection. In the course of the research, it was found that the implementation of the technological process in the presence of irreversible changes in the structure of the material of the restored part is not advisable, since irreversible changes in the material will lead to destruction during further operation, or at the stage of final control.

In order to determine and identify crack-like defects, a number of non-destructive testing methods are used, such as: ultrasonic (GOST 23667-85) device model UT2008, magnetic particle (GOST 21105-87), etc. The disadvantage of these methods is the low sensitivity threshold. The use of these methods makes it possible to detect crack-like defects from 1 mm in depth or more, which is already a relatively late stage of crack development.

In practice, a method for detecting fatigue microcracks is used, which consists in applying a metal film (for example, aluminum) to the structure. By the formation of local dark zones on the surface of the film (or by the discontinuity of the film) after loading, the appearance of microcracks in the test metal is recorded (RU 2390753 C1, G01N 3/32, 27.05.2010). The disadvantages of this method are the need for a film on the structure during the entire period of its operation, as well as the need to use special magnifying equipment to identify the results obtained.

A known method for determining the cyclic strength of the metal of structures, which consists in cyclic loading of a local area of the metal using an indenter and simultaneous magnetization and measurement of magnetization in the impact zone of the indenter (RU 2122721 C1, G01N 3/32, 27.11.1998). The disadvantage of this method is that the value of the magnetization of the metal is an indicator that reacts to the accumulation of damage in the metal, and cannot fix the moment of formation of microcracks in the metal.

The essence of the eddy current method lies in the analysis of the interaction of an external electromagnetic field with the electromagnetic field of eddy currents generated by the excitation coil in the electrically conductive structure of the blade.

The electromagnetic field of the eddy EMF acts on the coils of the converter, inducing an EMF in them or changing their electrical impedance. This effect allows you to get information about the properties of the object and the position of the transducer relative to it.

Thus, the parameters of the blade affect the electromotive force of the transducer, i.e. information, multi-parameter. This points to the advantage of implementing eddy current methods (ECM). Therefore, in this way, it is possible to obtain satisfactory diagnostic results even with high-speed movement of objects (Figure 7).

Information comes in the form of electrical signals. Wide range of non-contact capabilities and high performance determine the automation of eddy current testing. The signals of the eddy current transducer are practically not affected by pressure, humidity and pollution of the gaseous medium, radioactive radiation, contamination of the surface of the test object with non-conductive substances.

The metal magnetic memory method is one of the new diagnostic methods. The metal magnetic memory method (MMM), which reflects the structural and technological heredity of the product, and the corresponding control devices that do not require preparatory work, make it possible to provide 100% express control of the entire surface and identify components and parts on which it is not advisable to perform the recovery technology [64, 65].

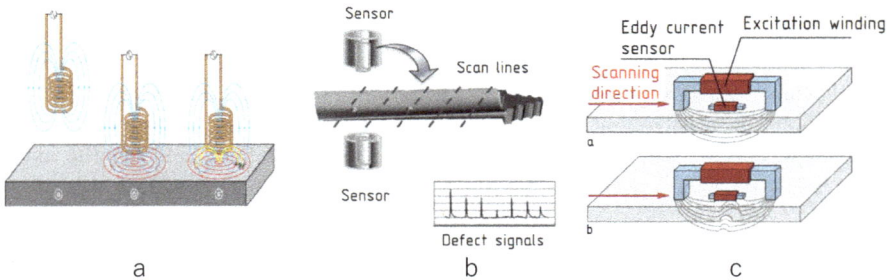

a - is an alternating current that passes through the coil at a selected frequency and creates a magnetic field; b - when a coil is placed near an object made of an electrically conductive material, eddy currents are excited in the object; c - if there is a defect in the object, it prevents the circulation of eddy currents and the magnetic connection is broken. Changes in the impedance of the coils indicate a defect

Figure 7. The principle of operation of eddy current flaw detectors

Monitoring the state of the metal using the MMM method on units and parts of power equipment before and after performing restoration repair technologies on them makes it possible to assess the quality of these technologies, structural changes, and the level and distribution of residual stresses [66, 67].

The method of metal magnetic memory is the most promising non-destructive testing method for practice in assessing the actual stress-strain state and mechanical properties of the metal. The development of this method will allow, simultaneously with diagnostics, to establish the causes of the formation of zones of stress concentration and damage.

To quantify the level of stress concentration (sources of damage) during scanning along the surface of the test object, the gradient of the normal (H_y) and/or tangential (H_x) components of the SMPR is determined [68]:

$$K = |\Delta H| / \Delta x, \text{ при } \Delta x \rightarrow 0 \ K = dH / dx,$$

where Δx is the distance between adjacent control points.

In some cases, when monitoring the stress-strain state (SSS) of equipment, the resulting SMPR is used, calculated from the results of measurements of three components and the gradient $|\Delta H| / \Delta x$.

Among the main calculated diagnostic parameters in the MMM method, the parameter m is used, which characterizes the limiting deformation capacity of the material:

$$m = K_{max} / K_{mid},$$

where K_{max} and K_{mid}, respectively, are the maximum and average values of the field gradient, which are determined during the control by the MMM method of equipment units of the same type.

Thus, analyzing the advantages and disadvantages of various methods of non-destructive testing of remanufactured turbine blades, we can draw the following conclusions:

1. The use of such methods of non-destructive testing as metal magnetic memory and eddy current with integrated, joint control of the blades will allow to achieve a complete diagnosis of the material in order to determine the maximum service life of the test object.
2. Using modern measuring instruments, it is impossible to determine and identify the fatigue indicators of the mechanical properties of the material (yield strength, τ, the magnitude and nature of stresses in the metal, etc.).
3. To solve this problem, it is necessary to use a systematic approach in assessing the structural-phase state of worn turbine blades, which determines their resource life.
4. In order to predict failures taking into account structural and phase changes, it is necessary to use probabilistic methods as a tool for assessing the reliability of CHP turbine blades and a package of applied programs.

To obtain reliable information about the properties of the metal structure of turbine blades, knowledge about the effect of structural-phase transformations on changes in the mechanical properties of the metal is necessary. It is possible to solve this problem by sharing the results of metallographic studies and flaw detection, with the integrated use of MMM and ECM methods.

1.5 Turbine Blade Reliability Assessment

The reliability of operation of a steam turbine is largely determined by the performance of the blade apparatus. Damage to rotor blades has the greatest impact on turbine downtime during refurbishment.

A high safety factor cannot be included in the calculation of the design of most steam turbine rotor blades, since this can lead to unjustified weighting not only of the blade itself, but also of the elements associated with it. On the other hand, failures of parts such as rotor blades (especially the last stage) lead to serious consequences [69, 70]. All this makes it necessary to improve the methods for calculating the structure during manufacture and to take into account more fully the factors that determine the stress-strain state of long blades.

The concept of predicting the resource of rotor parts for critical purposes should take into account operating conditions, the random nature of loading, degradation changes in turbine operation parameters and the effect of operating time on the mechanical properties of blade materials for various types of loading.

To predict the reliability of a turbine blade, it is necessary to investigate the dependencies that relate the acting stresses and steam temperature in the design section of the blade and disk and the temperature difference between the rim and hub of the disk with the parameters of the steam turbine.

For CHP turbine blades, these dependencies have the form [71-77]:

$$\begin{cases} \sigma_L = An^2 \\ t_L = BT_r^* + B_1 \end{cases};$$

(1)

For turbine discs:

$$\begin{cases} \sigma_D = A_1n^2 + A_2\Delta t \\ \Delta t = DT_r^* + D_1 \end{cases};$$

(2)

where σ_L and t_L are the total operating stress in the blade and the temperature of the blade;

σ_D - total operating stress on the disk;

Δt - temperature difference between the rim and the disc hub;

A, A_1, A_2, B, B_1, D, D_1 are constant coefficients for the considered turbine.

The use of dependencies (1) and (2) makes it possible to analyze the influence of various real loading factors on the durability of parts.

The durability of the working blade, τ_p according to the ultimate strength σ is estimated from the Larson-Miller dependence [78]:

$$\sigma = f(P_{L-M}), P_{L-M} = T(C + lg\tau_p),$$

(3)

where T is the steam temperature, K; C – constant (usually C=20);

The specific damage of the blade is calculated as the reciprocal of the durability τ_p (in hours) [79]:

$$\Pi_{0\tau} = \frac{1}{\tau_p};\tag{4}$$

The cyclic durability N_p of a disc is determined by a technique that includes the use of a schematized diagram of the deformation of the disc material and the Manson dependence [80].

One of the factors affecting the loading of parts of steam turbines of CHP is their mode of operation.

During long-term operation under thermomechanical loading, the mechanical properties and durability characteristics of materials of steam turbine parts change. At the same time, the strength properties of materials (ultimate strength σ_w, yield strength $\sigma_{0.2}$) remain practically unchanged, while plasticity characteristics (relative elongation δ, relative narrowing δ) decrease significantly, especially for disk materials [81, 82].

Based on the data of such tests, it is possible to derive dependencies characterizing the degradation changes in the operating parameters of the CHP steam turbine during operation:

$$\begin{cases} \overline{\Delta n} = k_1 \tau^{\alpha_1} \\ \overline{\Delta T_r^*} = k_2 \tau^{\alpha_2} \end{cases};\tag{5}$$

where Δn is the increment of the average rotor speeds;

$\overline{\Delta T_r^*}$ - average steam temperature increment;

τ - time between failures, h;

$k_1, k_2\, \alpha_1, \alpha_2$ are constants for the considered CHP steam turbine.

Since the values of the parameters (t^0, W, P, γ) of the CHP steam turbine increase during operation, the dynamic loads acting on the blade feather progressively increase. The process of dynamic loading is complicated by the fact that the acting forces and moments are distributed unevenly over the contact area of the blade. As a consequence, leading to the accumulation of the concentration of internal stresses in narrow sections of the blade structure.

After analyzing the standard methods for assessing the reliability of turbine blades, it was found that their essence boils down to theoretical calculations of the acting centrifugal forces, moments of inertia forces and stress state of the root section and blade airfoil in static mode. This approach does not take into account the dynamic moments that arise in the actual operating conditions of the CHP turbine. To improve the accuracy of forecasting, taking into account the physical and mechanical properties of rotor blades, the initial data for strength calculations must be corrected with the measurement results obtained by flaw detectors. The proposed concept will make it possible to take into account the actual changes in the mechanical characteristics of the material of rotor blades that have passed a certain period of operation. In turn, the process of changing the properties of the blade during operation will reflect the real conditions and the aggressive operating environment.

Thus, the task of improving the methods for predicting the resource and reliability of the turbines blades of CHP using instrumentation in synthesis with a probabilistic approach is very relevant.

Conclusion

Analyzing the prospect of being in demand among the population and enterprises of the machine-building industry, it was found that the most widely used are heat-extraction turbines T - 100/120-130, which provide the generation of electricity and heat.

1) It has been established that the main technological and operating parameters largely depend on the operating conditions of the turbine and the technical condition of the blade apparatus. The reliability of the blade apparatus is ensured and regulated by the system of diagnostics, maintenance and repair and restoration processes. This means that it is possible to make a theoretical assumption that the reliability of the turbine will depend on the quality indicators of the turbine blades, which are provided by a complex technological process of repair and restoration of worn surfaces of complex design geometry.

2) The analytical surfaces of rotor blades are a combination of linear, cylindrical and helical surfaces, which are quite simply formalized mathematically. The definition of a sculptural surface reflects the technological method of its formation, which must be taken into account during restoration work. However, the issue of methodological and technological support has not been sufficiently studied and there are significant gaps in the knowledge base for the restoration of the sculptural surface, taking into account the modification of the structure of the restored blade.

3) Previous researchers found that the choice of the shape of the blade root significantly affects the reduction of the centrifugal force acting on the blade and eliminates bending stresses in the disk rim. However, the conclusions were drawn under ideal conditions of factory production for blades that have not passed the operational period and do not have worn parts, the same shank. The presence of wear gaps in the tail part of the blade significantly change the force map and the principle of redistribution of the acting moments that provoke vibration loads and fatigue stresses along the design section of the blade airfoil.

4) Regularities of the distribution of dynamic and vibration loads over the entire surface of the length of the rotor blade have not been studied enough, as indicated by a large variety of defects of a different nature and in different zones of the blade airfoil. Models of formation of fatigue stresses in the structure of blades under cyclically acting constant and random loads have not been formed.

5) Having studied the technological features of laser recovery technologies, a difficult technical task is to synthesize the advantages of laser and plasma recovery technologies based on the use of highly concentrated energy sources. When developing innovative restoration technologies, it is necessary to clearly and unambiguously form the criteria and indicators of the quality of restoration of blades of complex geometry.

6) After analyzing the standard methods for assessing the reliability of turbine blades, it was found that their essence boils down to theoretical calculations of the acting centrifugal forces, moments of inertia forces and stress state of the root section and blade airfoil in static mode. To improve the accuracy of forecasting, taking into account the physical and mechanical properties of rotor blades, the initial data for strength calculations must be corrected with the measurement results obtained by flaw detectors. The success of the implementation of these tasks aimed at solving the scientific and technical problem of developing an energy-efficient laser-plasma technology for the restoration of substandard turbine blades, taking into account the actual changes in the mechanical characteristics of the material of the rotor blades that have passed a certain period of operation, largely depends on the adequacy of the mathematical description of the dynamic processes occurring in the turbine CHP.

Chapter 2. Mathematical Description of Dynamic Processes Occuring in the Turbine of the CHP

2.1 Study of Factors and Determination of Forces Acting in the Dynamic System "Steam-Blade-Turbine"

The dynamics of the turbine operation determines the technological requirements for it and affects the quality of the working process of its interaction with the contact medium. The flow of working steam directed to the turbine rotor and its blades has an effect on them, which depends on:

- steam pressure P, MPa;
- flow velocity at the entrance to the surface of the rotor and turbine blades and at the exit from them V, m/s;
- design shape of the blade surface ζ;
- angle of flow direction relative to this surface φ, rad;
- steam pressure difference in front of the blade and behind it ΔP, MPa.

The contact medium (degree of saturation of steam with moisture) of the CHP turbine has a destructive effect on it due to high operating temperatures T, pressure P and high-speed modes of operation, which, as a result, cause erosion, vibration of the turbine and deformation of the blades.

In the process of interaction of the rotor and turbine blades with the medium, resistance forces arise, leading to a violation of the equilibrium balance of forces and moments. The value of their values depends on the physical properties of the interaction medium, the geometry of the turbine blades and operating modes. All resistance forces can be represented as a system of normal and tangential components to the axis of rotation of the rotor.

The determination of the parameters and modes of operation of the blades is associated with the calculations of plates, shells and rod elements, which are the main bearing elements of the structures of turbomachines subjected to intense dynamic loads.

Studies on the calculation of shells, statics and dynamics of thin-walled structures are published in the works of Alexandrov A.B. [83], Bogdanovich A.E. [84], Bublik B.N. [85], Weinberg D. B. [86] and others. The functions describing the process of oscillations of the blades are given in the work of Schorr B. F. [72, 75]. In his research, Schorr B.F. focused on the factors affecting the emerging moments of forces, bending and shear deformation in the material.

In the works of Birger I.A., Schorr B.F. forced oscillations of the impellers of turbomachines are investigated using the method of initial parameters, but the method for calculating the amplitudes of the harmonics of the disturbing forces is not indicated [72 p. 95; 75 p. 200; 87].

The use of simplified calculation schemes based on the application of the rod theory [88] does not make it possible to take into account, for example, the effect of damping oscillation frequencies and the design parameters of the blades. In this case, the blades are considered as twisted inextensible and rigidly fixed in the disk rods of variable cross section, performing in-phase oscillations in one of the main bending planes.

In the works of Zhigalko Yu.P., Levin A.V. [89-91] solved the problems of shell dynamics under local influences on the basis of a general operator model of the dynamics of elastic systems using the method of expansion in terms of eigenmodes. Fundamental solutions of stationary and non-stationary problems are given, their structure is analyzed. The problem of vibrations of a thin shell of arbitrary shape with an attached rigid body is also considered. The analysis of the results of solving the problems of external and internal vibration damping of shells and plates under local influences is carried out.

Works by Leikin A.S. [92] are devoted to the description of the mechanism of occurrence of transverse vibrations of the blade apparatus, taking into account shroud links. To solve the problems of finding the dynamic characteristics of bar and thin-walled structures AI Golovanov, DV Berezhny, AV Pesoshin created software systems based on the finite element method.

When solving the equations of forced oscillations, asymptotic methods of nonlinear mechanics are used, but damping in the system is not taken into account [93]. In works [93 p. 113, 94] this problem is solved using the finite element method, but in [95, 96] damping is not taken into account, and in [93 p. 131] when constructing a finite element mathematical model of a blade ring, finite elements of the "tetrahedron" type are used, which do not allow to fully take into account all the design features, both of the airfoil and the blade root (Figure 8).

Ivanov V.P. [97] pointed out that in the theoretical calculation of the tone of oscillatory phenomena, the forms of vibrations are often distorted, which entails the dispersion of dynamic stresses over the elements of the step. Therefore, when designing and manufacturing a blade, for the purpose of high reliability, it is necessary to take into account the zones of stress formation.

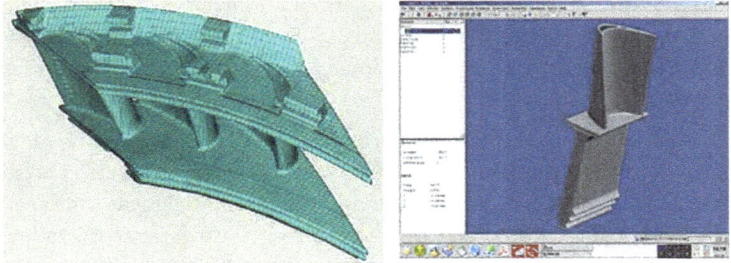

Figure 8. Turbine blade model

I.E. Zablotsky, Yu.A. Korosteleva, R.A. Shipova, K.N. Borishansky, E.V. Ureva, M.I. Kiseleva, A.N. Morozova, V.I. Pronyakin and other scientists [68, 100, 101 p. 43; 102 p. 34; 103 p. 58].

B.M. Abramova, E.L. Airapetova, B.A. Antufieva, I.I. Artobolevsky, F.Ya. Balitsky, L.Ya. Banakh, Yu.G. Barinova, M.D. Genkina, V.L. Dorofeeva, V.I. Erofeeva, O.I. Kosareva, B.V. Pavlova, P.P. Parkhomenko, A.G. Sokolova, O.F. [100]. Tishchenko, and other scientists. The method of experimental study of the frequencies of free vibrations of turbine blades is given in [101], however, the experimental data are less reliable, since in the process of the experiment there are always fatal system errors of the experimental setup itself and the measuring equipment.

Therefore, the nomenclature of existing methods for studying and predicting the failure-free operation of turbine unit elements allows:
- set the amplitude and frequency of oscillatory effects on the complex surface of the blades;
- to implement the primary adjustment of the operational modes of the blade apparatus in order to exclude the occurrence of resonance;
- establish zones of non-uniform dissipation of intermittent internal stresses under dynamic impact;
- neglect a large number of factors affecting the efficiency of the rotor operating modes.

The solution of the scientific and practical problem of the vibration reliability of the blade apparatus of turbines of thermal power plants requires the determination of the actual frequency characteristics and vibration stresses of the blades in real operating conditions, the study of the influence of regime operation on the reliability of the blades, and the justification of the optimal conditions for their operation [102, 103].

2.2   Improvement of the Mathematical Model of Dynamic Processes in the "Steam-Blade-Turbine" System

The process of interaction of the turbine rotor and its blades with the contact medium includes the following stages: steam entry to the blade, its movement along the blade, and exit from it. In this case, it is necessary to ensure that the steam jet does not hit the blade, but flows smoothly around it. Thus, it is possible to designate two contact zones of interaction between the surface of the turbine rotor and its blades with the contact medium: the zone of direct contact and the zone of steam removal from the blade. The stages of steam entry to the blade, steam movement along the blade is carried out in the zone of direct contact, and steam exit - in the zone of steam removal from the blade. Elementary forces attributable to the surface of the rotor and turbine blades are shown in Figure 9.

Let's consider the indicated stages in more detail. The resistance to rotation of the turbine with blades W includes the force of resistance to rotation of the turbine rotor in the contact medium P; steam friction force on the surface of the turbine rotor; force on the turbine blades, T; the force required to remove steam from the surface of the blades P_{vyn}:

$$\vec{W} = \vec{P} + \vec{\Delta} + \vec{T} + \vec{P}_{vyn} , \tag{6}$$

Where P is the sum of projections of normal pressure forces from the side of the medium onto the surface of the turbine rotor,

Δ - the sum of the projections of friction forces from the side of the contact medium on the surface of the turbine rotor,

T - the sum of the projections of the forces of resistance to the movement of the blades in the steam medium,

P_{vyn} - is the force required to carry steam away from the blade.

The total force on the blades T, located in the steam medium, is defined as the sum of the product of the forces on the blades in the steam medium and the blade angle φ:\

$$T = \sum_{i=1}^{n} T_i \cos \phi_i. \tag{7}$$

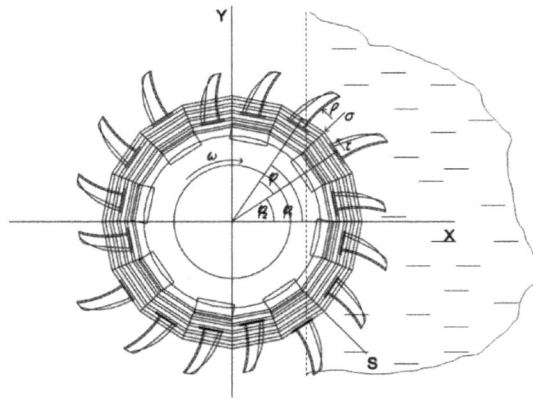

Figure 9. Scheme of elemental forces acting on the surface of the rotor and some turbine blades when interacting with the contact medium (front view)

The friction force of the contact medium on the surface of the turbine rotor can be found as the integral sum of the friction forces of the soil on the elementary areas of the turbine, directed tangentially to its surface

$$\Delta = \iint_S \tau \cos \phi_p \, dS, \tag{8}$$

where τ is the friction force of the contact medium, directed tangentially to the surface of the turbine rotor,

S is the contact surface of the rotor with the medium,

dS is the elementary area of contact between the medium and the surface of the turbine rotor.

The total value of the force of resistance to the movement of the turbine rotor from the forces of normal pressure of the contact medium on its surface is determined from the expression

$$P = \iint_S \sigma dS \cdot \cos \left(\overset{\wedge}{\sigma}, y \right), \tag{9}$$

where is the elementary normal pressure from the side of the contact medium on the side surface of the turbine rotor.

The force required to remove steam from the blade is determined from the expression

$$P_{vyn} = P_{sdv} - P_{ost}$$

,

$$(10)$$

where P_{sdv} is the force that shifts steam from the surface of the blade,

P_{ost} -residual force of the contact medium acting on the surface of the blade.
To determine the sum of projections of forces on the blades, a design scheme of forces has been developed (Figure 10) [103, 104].

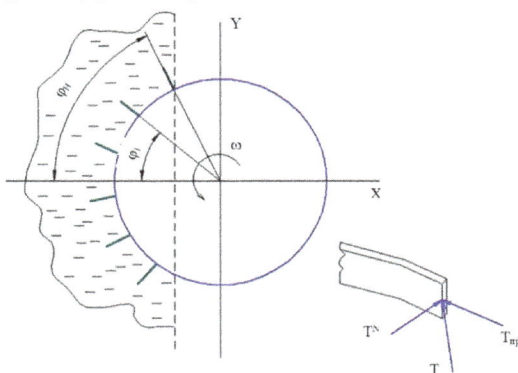

Figure 10. Scheme of forces acting on the blades (front view)

The movement of the blade is accompanied by a change in the angle of attack of the medium. As a result, the reaction force of the blade also changes. When the blade moves in the contact medium, a force of resistance to the movement of the blade T arises. It includes the normal T^N and tangential T_{tr} components.

The total normal force on the blades T^N, which are in the contact medium, is defined as the sum of the product of the forces on the blades and the angle of installation of the blade φ_i:

$$T^N = \sum_{i=1}^{n} T_i^N \cos \varphi_i$$

,

$$(11)$$

where T_i is the resistance force that falls on the blade when it comes into contact with steam.

27

$$T_i^N = p \cdot b_0 \cdot h_i \cdot cos\,\varphi_i \,.$$
(12)

The normal component of the displacement resistance force is determined by the dependence

$$T^N = \sum_{i=1}^{n} p \cdot b_0 \cdot h_i \cdot \cos\varphi_i \,,$$
(13)

where p is the pressure of the contact medium on the blade;

b_0 - blade width;

h_i - steam thickness between the blades;

n is the number of blades.

Because blades are set by pitch angle, can be written

$$\varphi_i = \varphi_1 + (i-1) \cdot \varphi_0 \,,$$

where φ_1 is the angle of steam capture by the blade.

Expression (2.8) takes the form:

$$T^N = \frac{p \cdot b_0}{2} \cdot \varphi_0 \frac{\sum_{i=1}^{n} \sin 2\varphi_i}{\omega} = \frac{p \cdot b_0}{2} \cdot \frac{\varphi}{\omega} \sum_{i=1}^{n} \sin\left[2\varphi_1 + (i-1)\cdot 2\varphi_0\right] \,.$$
(14)

We use the well-known formula [72 p. 13], taking into account which from expression (14) we have

$$
\begin{aligned}
T^N &= \frac{p \cdot b_0}{2} \cdot \frac{\varphi_0}{\omega} \cdot \frac{\sin\left[2\varphi_1 + (n-1)\cdot\varphi_0\right] \cdot \sin(n\cdot\varphi_0)}{\sin\varphi_0} = \\
&= \frac{(\omega \cdot R)^2 \cdot \gamma \cdot b_0}{2} \cdot \frac{\varphi_0}{\omega} \cdot \frac{\sin\left[2\varphi_1 + (n-1)\cdot\varphi_0\right] \cdot \sin(n\cdot\varphi_0)}{\sin\varphi_0} = \\
&= \frac{\gamma \cdot b_0 \cdot \varphi_0 \cdot R^2 \cdot \omega}{2} \cdot \frac{\sin\left[2\varphi_1 + (n-1)\cdot\varphi_0\right] \cdot \sin(n\cdot\varphi_0)}{\sin\varphi_0} \,.
\end{aligned}
$$
(15)

where R - is the radius of installation of the blade on the surface of the turbine rotor; ω - angular speed of rotation of the turbine rotor.

If we assume that the turbine has K rows of blades, and taking into account that formula (15) determines the resistance force per row of turbine blades, then we finally have an expression for determining the magnitude of the resistance force to rotation of the turbine blades in the contact medium [104 p. 35]:

$$T^N = 2\sum_{i=1}^{K} T_i = \frac{\gamma \cdot b_0 \cdot \varphi_0}{\omega \cdot \sin \varphi_0} \cdot \sum_{i=1}^{K} R_i^2 \cdot \sin\left[2 \cdot \varphi_1 + (n_i - 1) \cdot \varphi_0\right] \cdot \sin(n \cdot \varphi_0), \tag{16}$$

where R_i is the radius of installation of the blades on the rotor;

φ_{1i} - angle of rotation of the blade;

n_i - the number of blades of the i-th row.

Accordingly, the friction force directed tangentially to the blade surface is defined as:

$$T_{mp} = T^N \cdot f, \tag{17}$$

where f is the coefficient of external friction.

The force of friction of the contact medium on the surface of the turbine rotor Δ can be found as the integral sum of the friction forces on the elementary areas of the turbine rotor directed tangentially to its surface (Figure 11).

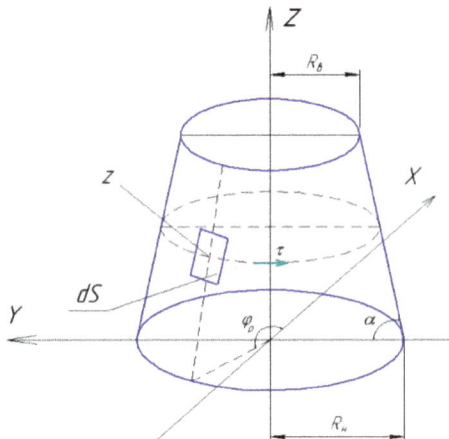

Figure 11. Scheme of forces acting from the contact medium on the side surface of the stepped turbine rotor (top view)

The friction force of the contact medium on the side surface of the turbine rotor, according to the design scheme shown in Figure 11, is determined as follows:

$$\Delta = 2\iint_S \tau \cos\varphi_p \, dS. \quad (18)$$

We express the shear stress in terms of the normal one:

$$\tau = \sigma \cdot f = f(H - z)\gamma \cdot \xi \qquad (19)$$

where f, ξ are the coefficients of friction and lateral pressure,

H is the length of the turbine rotor;

z - coordinate of an arbitrary point on the side surface of the turbine rotor

$$z = (R - \sqrt{x^2 + y^2}) \cdot tg\alpha, \qquad (20)$$

where α is the angle of inclination of the generatrix of the rotor to the horizontal surface,

R - is the radius of the turbine rotor.

The area of the elementary area of the lateral surface of the turbine rotor

$$dS = \frac{dx \cdot dy}{\cos\alpha}, \qquad (21)$$

$$\cos\varphi_p = -\frac{x}{\sqrt{x^2 + y^2}}. \qquad (22)$$

Taking into account expressions (16) - (22), the integral (14) takes the form:

$$
\frac{\Delta}{2} = -\iint f\xi\gamma\left(H - tg\alpha\left(R_{_H} - \sqrt{x^2 + y^2}\right)\right)\frac{xdxdy}{\sqrt{x^2 + y^2} \; \cos\alpha}\right) =
$$

$$
= -\frac{f\xi\gamma}{\cos\alpha}\iint\left[(H - R_{_H}tg\alpha)\frac{x}{\sqrt{x^2 + y^2}} + xtg\alpha\right]dxdy
\qquad (23)
$$

where $R_{_H}$ is the maximum radius of installation of blades on the turbine rotor.

Taking into account the area of integration shown in Figure 12, from the last expression we have:

$$\frac{\Delta}{2} = -\frac{f\xi\gamma}{\cos\alpha}\int dy \int_{x_3}^{x_1,x_2}\left[(H - R_n tg\alpha)\frac{x}{\sqrt{x^2 + y^2}} + xtg\alpha\right]dx, \qquad (24)$$

where x_1, x_2, x_3 are the equations of the lines limiting the integration area (Figure 12).

$$x_1 = x_0 = -(a - 0{,}5b), \qquad (25)$$

$$x_2 = -\sqrt{R_e^2 - y^2}, \qquad (26)$$

$$x_3 = -\sqrt{R_n^2 - y^2}, \qquad (27)$$

where a is the distance from the axis of rotation of the turbine rotor to its housing,

b - turbine casing width,

R_e - minimum radius of installation of blades on the turbine rotor

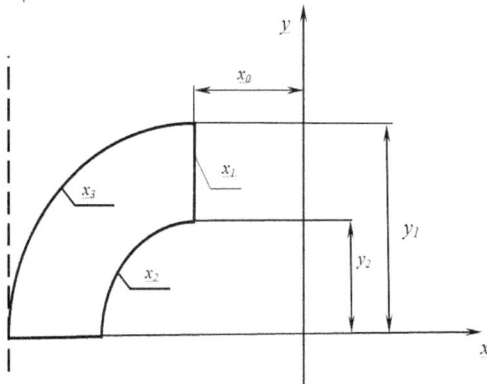

Figure 12. Integration area

On the right side of the double integral (28), we perform integration over the variable x:

$$\frac{\Delta}{2} = \frac{x\xi\gamma_{e}}{\cos\alpha} \int_{0}^{y_2} dy \left[(H - R_{_H}tg\alpha)\sqrt{x^2 + y^2} + \frac{x^2}{2}tg\alpha \right]_{}^{x_2} + \tag{28}$$

$$+ \frac{f\xi\gamma}{\cos\alpha} \int_{y_2}^{y_1} dy \left[(H - R_{_H}tg\alpha)\sqrt{x^2 + y^2} + \frac{x^2}{2}tg\alpha \right]_{}^{x_1} +$$

$$+ \frac{f\xi\gamma}{\cos\alpha} \int_{0}^{y_1} dy \left[(H - R_{_H}tg\alpha)\sqrt{x^2 + y^2} + \frac{x^2}{2}tg\alpha \right]_{}^{x_3} =$$

$$= \frac{f\xi\gamma}{\cos\alpha} \int_{0}^{y_2} \left[(H - R_{_H}tg\alpha) \cdot R_{_e} + (R_{_e}^2 - y^2)\frac{tg\alpha}{2} \right] dy +$$

$$+ \frac{f\xi\gamma}{\cos\alpha} \int_{y_2}^{y_1} \left[(H - R_{_H}tg\alpha)\sqrt{x_0^2 + y^2} + \frac{x_0^2}{2}tg\alpha \right] dy +$$

$$+ \frac{f\xi\gamma}{\cos\alpha} \int_{0}^{y_1} \left[(H - R_{_H}tg\alpha)R_{_H} + (R_{_H}^2 - y^2)\frac{tg\alpha}{2} \right] dy$$

where

$$y_1 = \sqrt{R_{_H}^2 - x_0^2}\,, \tag{29}$$

$$y_2 = \sqrt{R_{_e}^2 - x_0^2}\,. \tag{30}$$

Integrating expression (30) over the variable, we finally have:

$$\Delta = \frac{x\xi\gamma}{\cos\alpha} \left[(H - R_{_H}tg\alpha)R_{_H} \cdot y_1 + \left(R_{_H}^2 y_1 - \frac{y_1^3}{3} \right)tg\alpha \right] - \tag{31}$$

$$- \frac{f\xi\gamma}{\cos\alpha} \left[2 \cdot R_{_b} \cdot y_2(H - R_{_H}tg\alpha) + \left(R_{_e}^2 \cdot y_2 - \frac{y_2^3}{3} \right)tg\alpha \right] -$$

$$- \frac{f\xi\gamma}{\cos\alpha} \left[(H - R_{_H}tg\alpha)(y_1 R_{_H} - y_2 R_{_e}) + \frac{x_0^2}{2}\ln\left|\frac{y_1 + R_{_H}}{y_2 + R_{_e}}\right| \frac{x_0^2}{2}(y_1 - y_2)tg\alpha \right].$$

Let us determine the sum of projections of normal pressure forces from the side of the contact medium onto the surface of the turbine rotor

The total resistance to the movement of the turbine rotor from the forces of normal pressure of the contact medium:

$$P = \iint_S \sigma \, dS \cdot \cos\left(\overset{\wedge}{\sigma, y}\right)$$

where $\sigma -$ is the elementary normal pressure from the side of the contact medium on the surface of the turbine rotor

$$\sigma = (H - z)\gamma \cdot \xi, \tag{32}$$

Where γ is the density of the contact medium,

ξ - side pressure coefficient,

$\cos\left(\overset{\wedge}{\sigma, y}\right)$ is the slope of the normal pressure vector

$$\cos\left(\overset{\wedge}{\sigma, y}\right) = \frac{\dfrac{\partial F}{\partial y}}{\sqrt{\left(\dfrac{\partial F}{\partial x}\right)^2 + \left(\dfrac{\partial F}{\partial y}\right)^2 + \left(\dfrac{\partial F}{\partial z}\right)^2}}, \tag{33}$$

where $F = F(x, y, z) = 0$ is the implicit surface equation:

$$F = z + tg\alpha\left(\sqrt{x^2 + y^2} - R_H\right) = 0. \tag{34}$$

We find the partial derivatives of the last equation and substitute them into expression (34):

$$\cos\left(\overset{\wedge}{\sigma, y}\right) = \frac{y \sin\alpha}{\sqrt{x^2 + y^2}}. \tag{35}$$

Taking into account expressions (32) - (35), equation takes the form:

$$P = 2\gamma\xi \, tg\alpha \iint_{(s)} (H - z) \cdot \frac{y}{\sqrt{x^2 + y^2}} \, dxdy = 2\gamma\xi \cdot J, \tag{36}$$

where

$$J = \iint_{(s)} (H - z) \cdot \frac{y \, dxdy}{\sqrt{x^2 + y^2}}. \tag{37}$$

Taking into account dependence (37)

$$J = \iint\limits_{(s)} \left[H - \left(R_H - \sqrt{x^2 + y^2} \right) tg\alpha \right] \cdot \frac{y\,dx\,dy}{\sqrt{x^2 + y^2}} = \iint\limits_{(s)} \left[(H - R_H tg\alpha)\frac{y}{\sqrt{x^2 + y^2}} + ytg\alpha \right] dx\,dy. \quad (38)$$

We integrate the resulting expression, taking into account the area of integration (Figure 12):

$$J = \int dx \left[(H - R_H \cdot tg\alpha)\sqrt{x^2 + y^2} + \frac{y^2}{2}tg\alpha \right]_{y_2\,y_3}^{y_1}, \quad (39)$$

where

$$y_1 = \sqrt{R_H^2 - x^2},$$
$$y_2 = 0,$$
$$y_3 = \sqrt{R_в^2 - x_2}. \quad (40)$$

Taking into account the integration boundaries (40), we obtain:

$$J = \int\limits_{-R_г}^{x_0} \left[(H - R_н tg\alpha)R_н + \frac{(R_н^2 - x^2)}{2}tg\alpha \right] dx - \int\limits_{-R_в}^{-R_в}(H - R_н tg\alpha)x\,dx -$$

$$- \int\limits_{-R_в}^{x_0} \left[(H - R_н tg\alpha)R_в + \frac{(R_в^2 - x^2)}{2}tg\alpha \right] dx =$$

$$= \left[(H - R_н tg\alpha)R_н(x_0 + R_н) + \frac{R_н^2(x_0 + R_н) - (x_0^3 + R_н^3)/3}{2}tg\alpha \right] -$$

$$- (H - R_н tg\alpha)\frac{(R_в^2 - R_н^2)}{2} -$$

$$- \left[(H - R_н tg\alpha)R_в(x_0 + R_в) + \frac{R_в^2(x_0 + R_в) - (x_0^3 + R_в^3)/3}{2}tg\alpha \right] =$$

$$= (x_0 + R_н)\left[(H - R_н tg\alpha)R_н + (2R_н^2 - x_0^2 + x_0 R_н)\frac{tg\alpha}{6} \right] -$$

$$- (x_0 + R_в)\left[(H - R_н tg\alpha)R_в + (2R_в^2 - x_0^2 + x_0 R_в)\frac{tg\alpha}{6} \right] - (H - R_н tg\alpha)\frac{(R_в^2 - R_в^2)}{2}.$$

Substituting the obtained value of the integral into expression (36), we finally have:

$$P = 2\gamma\xi \, tg\alpha \left\{ (x_0 + R_{_H}) \left[(H - R_{_H} tg\alpha)R_{_H} + \left(2R_{_H}^2 - x_0^2 + x_0 R_{_H}\right)\frac{tg\alpha}{6} \right] - \right.$$
$$- (x_0 + R_{_6}) \left[(H - R_{_H} tg\alpha)R_{_6} + \left(2R_{_6}^2 - x_0^2 + x_0 R_{_6}\right)\frac{tg\alpha}{6} \right] - \tag{41}$$
$$\left. - (H - R_{_H} tg\alpha)\frac{\left(R_{_H}^2 - R_{_6}^2\right)}{2} \right\}.$$

To determine the magnitude of the force required to remove steam from the surface of the turbine rotor blade, it is necessary to take into account the centrifugal force of inertia $F_w = m \cdot \omega^2 \cdot R$, the normal reaction of the blade surface T^N and the friction force $T_{mp} = f \cdot T^N$.

The force that shifts steam from the surface of the blade is:

$$P_{c\partial s} = m\omega^2 R \cdot \sin\varphi . \tag{42}$$

Residual force of the contact medium on the surface of the blade:

$$P_{ocm} = \left(\frac{\gamma \cdot b_0 \cdot \varphi_0 \cdot R^2 \cdot \omega}{2} \cdot \frac{\sin[2\varphi_1 + (n-1) \cdot \varphi_0] \cdot \sin(n \cdot \varphi_0)}{\sin\varphi_0} + \frac{\gamma \cdot b_0 \cdot R^2}{\omega} \cdot (1 + f) \right) \cdot \sin\varphi . \tag{43}$$

On the rotor of the turbine unit, on the one hand, the torque that occurs during the rotational movement of the rotor blades when steam is supplied to the flow path of the turbine, and on the other hand, the moment of forces of resistance to the rotation of the turbine rotor from the generator side.

In the steady state of operation, the rotor rotates uniformly at a constant speed, which is possible only if the moments of the driving forces on the turbine blades M_T and the resistance forces of the M_G are equal:

$$M_{T0} - M_{\Gamma 0} = 0. \tag{44}$$

Index zero corresponds to steady motion.

During the operation of the turbine, the equality of the moments of forces (44) may be violated due to the modes of its operation (for example, when the steam pressure changes, its flow rate in the flow path of the turbine or when the electrical resistance of devices and units connected to the generator changes).

Unsteady motion can be described by Euler's theorem on the change in angular momentum. In this case, the time derivative of the main moment of the system movement relative to the rotor axis is equal to the main moment of external forces relative to this axis [103 p.115]:

$$\frac{d}{dt}(J\omega) = M_{\text{T}} - M_{\Gamma}, \tag{45}$$

where J is the moment of inertia of the turbine rotor; ω - angular speed of rotation of the rotor.

$$M_{\text{T}} = M_{\text{T0}} + \Delta M_{\text{T}}, \tag{46}$$
$$M_{\Gamma} = M_{G0} + \Delta M_{\Gamma}, \tag{47}$$

where ΔM_{T} and ΔM_G change in the moments of the driving forces on the turbine blades and the forces of resistance to rotation, respectively.

From expressions (44) and (47) we obtain

$$J\frac{dw}{dt} = \Delta M_{\text{T}} - \Delta M_G. \tag{48}$$

The moment of driving forces on the turbine blades can be considered as a function of several variables:

$$M_{\text{T}} = f(P_1, T_1, P_{\text{K}}, k, \omega), \tag{49}$$

where P_1, T_1 are steam parameters in the turbine,

P_k - steam pressure in the turbine condenser,

k is the opening value of the turbine control valves

The moment of resistance forces on the shaft of the generator M_G depends on the speed of rotation of the shaft ω and the total electrical resistance connected to the generator and depending on time t. Therefore, the change in the forces of resistance to rotation Γ can be written as ΔM_G

$$\Delta M_G = \Delta M_G(\omega) + \Delta M_G(t). \tag{50}$$

If we assume that the isentropic enthalpy drop (heat drop) in the process of steam expansion in the flow path of the turbine is not changed, that is, the steam parameters P_1, T_1, P_k are constant, then the moment of driving forces on the turbine blades M_T from expression (50) is a function of two variables:

$$M_T = f(k, \omega). \tag{51}$$

Let us expand the functions M_T and M_G in a Taylor series, leaving only the terms of the series containing deviations not higher than the first

$$\Delta M_T = \left(\frac{\partial M_T}{\partial k}\right)_0 \cdot \Delta k + \left(\frac{\partial M_T}{\partial \omega}\right)_0 \cdot \Delta \omega, \tag{52}$$

$$\Delta M_G = \left(\frac{\partial M_G}{\partial \omega}\right)_0 \cdot \Delta \omega + \Delta M_G(t). \tag{53}$$

Substituting the last two dependencies into (48), we obtain the equation for the rotor of the turbine unit, taking into account small deviations of the independent variables from the steady state

$$J\frac{d(\Delta\omega)}{dt} = \left(\frac{\partial M_T}{\partial k}\right)_0 \cdot \Delta k + \left(\frac{\partial M_T}{\partial \omega}\right)_0 \cdot \Delta \omega - \left(\frac{\partial M_\Gamma}{\partial \omega}\right)_0 \cdot \Delta \omega + \Delta M_\Gamma(t). \tag{54}$$

Denote:

$$\frac{\Delta\omega}{\omega_0} = \varphi; \tag{55}$$

$$\frac{k}{k_{max}} = \mu; \tag{56}$$

$$\frac{\Delta M_\Gamma(t)}{M_{\Gamma max}} = \lambda, \tag{57}$$

where ω_0 is the nominal value of the angular velocity of rotation of the turbine rotor;

k_{max} is the maximum displacement of the control valve of the turbine unit, corresponding to the change in the load of the turbine unit from idle to maximum;

M_{Gmax} - maximum moment of resistance forces on the generator shaft.

Let us write equation (54) in relative values of independent variables:

$$J\omega_0 \frac{d\varphi}{dt} = \left(\frac{\partial M_T}{\partial k}\right)_0 \cdot k_{max}\mu + \left(\frac{\partial M_T}{\partial \omega}\right)_0 \cdot \omega_0\varphi - \left(\frac{\partial M_G}{\partial \omega}\right)_0 \cdot \omega_0\varphi + \frac{\Delta M_G(t)}{\Delta M_{Gmax}} \cdot M_{\Gamma max}. \tag{58}$$

Let's separate the variables:

$$J\omega_0 \frac{d\varphi}{dt} + \left[\left(\frac{\partial M_G}{\partial \omega}\right)_0 - \left(\frac{\partial M_T}{\partial \omega}\right)_0\right] \cdot \omega_0\varphi = \left(\frac{\partial M_T}{\partial k}\right)_0 \cdot k_{max}\mu - M_{Gmax}\lambda. \tag{59}$$

Let's agree that $J\omega_0$; $\left[\left(\frac{\partial M_G}{\partial \omega}\right)_0 - \left(\frac{\partial M_T}{\partial \omega}\right)_0\right] \cdot \omega_0$; $\left(\frac{\partial M_T}{\partial k}\right)_0 \cdot k_{max}$; M_{Gmax} constant coefficients for independent variables

Denote

$$N = \frac{J}{\left[\frac{\partial M_\Gamma}{\partial \omega} - \frac{\partial M_T}{\partial \omega}\right]}; \tag{60}$$

$$N_\mu = \frac{J\omega_0}{\left(\frac{\partial M_T}{\partial k}\right)_0 k_{max}}; \tag{61}$$

$$N_\lambda = \frac{J\omega_0}{M_{\Gamma max}}. \tag{62}$$

Then expression (59) will take the form:

$$\frac{d\varphi}{dt} + \frac{\varphi}{N} = \frac{\mu}{N_\mu} - \frac{\lambda}{N_\lambda}, \tag{63}$$

or

$$N\frac{d\varphi}{dt} + \varphi = \beta_1\mu - \beta_2\lambda, \tag{64}$$

where N is the dynamic constant of the rotor;

β - dimensionless coefficients characterizing the static properties of the system.

$$\beta_1 = \frac{N}{N_\mu}; \tag{65}$$

$$\beta_2 = \frac{N}{N_\lambda}, \tag{66}$$

$$N = \frac{J}{\frac{N_0}{\omega_0^2} + \frac{N_0}{\omega_0^2}} = \frac{J\omega_0^2}{2N_0}. \tag{67}$$

For a linear characteristic of the turbine control elements:

$$N_0 = M_0 \cdot \omega_0 .$$

The strength indicator for evaluating the resource of the blades is the value of the safety margin, which is represented by the ratio of the ultimate stress σ_{limit} to the highest value of the total stress $K=\sigma_{limit}/\sigma_{\sum max}$.

2.3   Improvement of the Mathematical Model of Dynamic Processes in the "Steam-Blade-Turbine" System

From the developed mathematical model of the dynamic processes of the CHP turbine, it has been established that the root section of the tail part of the blade experiences the least load. Turbine operation experience and analysis of blade defect reports showed that safety factors range from 1.8 to 2.3.

In order to improve the mathematical model of the dynamic processes of the turbine, the blade, which perceives the main types of loads, is conditionally divided by sections (i = 1...6) into five identical parts. The first section is combined with the root, and the sixth - with the peripheral section of the blade. The division of the blade into parts will help to investigate the dynamic forces acting on the elements of the turbine in transient modes and modes that provide the most efficient turbine power. Blade profile parameters for six sections are given in Table 4.

Table 4. Blade profile parameters for six sections

Geometric profile settings	Radius of the investigated section, m					
	0.590	0.635	0.675	0.715	0.755	0.795
$b \cdot 10^3$, m	0.071	0.072	0.073	0.074	0.075	0.077
$\delta \cdot 10^3$, m	0.015	0.013	0.012	0.010	0.008	0.007
$h \cdot 10^3$, m	0,035	0.030	0.026	0.023	0.019	0.016
β, grad	14.365	20.895	27.500	33.345	38.767	44.247
Note - b - profile chord; δ – profile thickness; h - profile deflection; β is the angle between the axis of rotation of turbomachines and the axis of minimum rigidity.						

For each design section, we determine the following geometric characteristics, taking into account erosion-corrosion changes:

a) cross-sectional area of the new blade $F = 0.7\ b\delta$;

b) the cross-sectional area of the blade, taking into account erosion-corrosion changes:

$$F_{er} = 0{,}7d_p(\delta_l' + b_i' \cdot R_a),$$

where d_p is the diameter of the area with erosion-corrosion changes, mm;

δ_l' - profile thickness, in the area with erosion-corrosion changes, mm;

b_i' - depth of the blade profile in the area with erosion-corrosion changes, mm;

R_a - surface roughness of the area with erosion-corrosion changes, microns.

Thus, the cross-sectional area of a worn blade is determined by the formula:

$$\sum F_{total} = F + F_{er} \cdot n_{plot}$$

where n_{plot} is the number of erosion-corrosion areas on the considered blade.

c) coordinates of the centers of mass of the section $b_c = 0.43b$; $h_c = 0.76h$.

The obtained values of F, b_c, h_c are entered in table 5.

Table 5. Estimated values

Geometric	Section number					
profile settings	1	2	3	4	5	6
R, m	0.590	0.635	0.675	0.715	0.755	0.795
$F \cdot 10^4$, m	7.077	6.365	5.695	5.139	4.628	4.149
$b_c \cdot 10^3$, m	30.574	30.617	30.660	31.134	31.907	33.115
$h_c \cdot 10^3$, m	25.090	22.450	19.810	17.171	14.531	11.891
ηA, m	0.025	0.022	0.020	0.017	0.015	0.012
ηC, m	-0.015	-0.013	-0.012	-0.010	-0.009	-0.007
ηB, m	0.025	0.022	0.020	0.017	0.015	0.012
ξA, m	-0.031	-0.031	-0.031	-0.031	-0.032	-0.033
ξC, m	0.041	0.041	0.041	0.041	0.042	0.044
$J_\eta \cdot 10^9$, m^4	376.593	302.268	236.617	181.974	135.297	95.927
$J_\xi \cdot 10^8$, m^4	192.686	173.769	155.918	145.076	137.224	132.475

The central moments of inertia J_η and J_ξ for each section of a new blade are determined by the formulas:

$$J_\eta = 0.0377 b^3 \delta;$$

$$J_\xi = 0{,}041 b \delta^2 \left[1 + \left(\frac{h}{\delta} \right)^2 \right]. \tag{68}$$

Mathematical description of the change in the values of the central moments of inertia J_η and J_ξ for each section, studied taking into account erosion and corrosion changes in real operating conditions.

$$J'_\eta = 0{,}377 (b + d_p)^3 \cdot (\delta'_l + b'_i \cdot R_a)$$

$$J'_\xi = 0{,}041 (b + d_p) \cdot (\delta'_l + b'_i \cdot R_a)^2 \cdot \left[1 + \left(\frac{h}{\delta'_l + b'_i \cdot R_a} \right)^2 \right]$$

where h is the deflection of the blade profile.

The results of the calculation are entered in table 6.

The following equalities deserve special attention: the forces acting from the center of mass, in sections i = 6, 5, and the tensile stresses caused by them:

$$P_{ji} = A(R_5^2 - R_i^2) \left(\frac{k+1}{2k} \right) \sum F_{total\ i} \tag{69}$$

$$\sigma_{pi} = \frac{P_{ji}}{\sum F_{total\ i}} \tag{70}$$

where $k = \dfrac{\sum F_{total\ i}}{F_i}$ change in the cross-sectional area at the calculated radius Ri ;

F_i - cross-sectional area in the considered radius;

A - is a constant factor.

The presence of a bandage will cause additional stresses in the working blade:

$$\Delta\sigma_\delta = \frac{\rho_\delta \cdot \omega^2 \cdot f_\delta \cdot t_\delta \cdot (r_\kappa + R_\delta)}{\sum F_{o\delta u}} = \frac{7{,}750 \cdot 314^2 \cdot 4.94 \cdot 10^{-5} \cdot 0.0168 \cdot (0.54 + 0.563)}{2.119 \cdot 10^{-4}} = 3.31 \text{МПа}$$

where f_b is the cross-sectional area of the bandage, $f_b = 4.94 \cdot 10^{-5}$ m^2;

t_b is the pitch of the blades along the circumference of the shroud, $t_b = 2\pi R_b/z = 0.0168$ m;

r_k = 0.54 m - root radius;

$\sum F_{total}$ = 0.0002119 m^2 - sectional area of a worn blade.

Let us describe mathematically the bending of the working blade in the event of a shroud band break. Let us determine the bending stresses acting on the working blade along the entire length. The action of the working fluid on the blade creates a force that can be decomposed into circumferential P_u and axial P_z components, Figure 13 [104 p. 36].

Table 6. Estimated data

Geometric profile parameters	Radius of the investigated section					
	1	**2**	**3**	**4**	**5**	**6**
R_i, m	0.590	0.635	0.675	0.715	0.755	0.795
$R_5^2 - R_i^2$, m^2	0.274	0.225	0.174	0.119	0.061	0.000
F_i 10^4,m^2	7.077	6.365	5.695	5.139	4.628	4.149
k	1.706	1.534	1.373	1.239	1.115	1.000
$\dfrac{k+1}{2k}$	0.793	0.826	0.864	0.904	0.948	1.000
P_{ji}10^{-3},H	16.347	27.057	55.801	86.381	119.697	155.339
σ_{pi},MPa	32.453	58.465	108.584	151.686	188.068	219.489
N_{TB}, kW	79.65	89.67	101.2	124.204	181.954	252.691
J'_η•10^9, m^4	119.909	169.121	227.468	295.771	377.835	470.741
J'_ξ•10^9, m^4	165.594	171.530	181.345	194.898	217.211	240.858

The power spent on friction and ventilation N_{tv}.

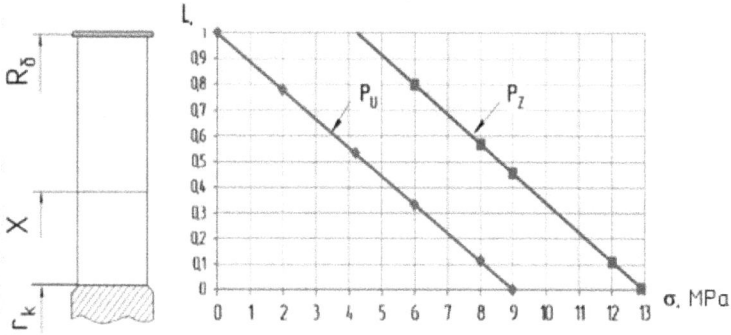

Figure 13. Graph of tensile stresses in the blade front

The axial component of the force acting on the blade:

$$P_z = \frac{G}{z} \cdot (C_1 \cdot \sin \alpha_1 - C_2 \cdot \sin \alpha_2) + (P_1 - P_2) \cdot t_2 \cdot l =$$
$$= \frac{20.845}{210} \cdot (319.267 \cdot \sin 12° - 57.185 \cdot \sin(-70.28)°) + (1{,}294 - 1{,}220)$$
$$\cdot 0.0182 \cdot 0.022 = 30.66 \text{ H}$$

The circumferential component of the force acting on the blade:

$$P_{\text{и}} = \frac{G}{z}(C_1 \cdot \cos \alpha_1 - C_2 \cdot \cos \alpha_2) =$$
$$= \frac{20.845}{210}(319.267 \cdot \cos 12° - 57.185 \cdot \cos(-70.28)°) == 29.08 \text{ H}$$

Forces acting on the blade:

$$P = \sqrt{P_z^2 + P_{\text{и}}^2} = \sqrt{30.66^2 + 29.08^2} = 42.26 \text{ H}$$

Bending moment:

$$M_{bend} = \frac{P \cdot l}{2}$$

The main quantitative indicators of the calculation for the bending of the working blade connected with the shroud are presented in Table 7.

Table 7. Main indicators

ζ	$I(\xi)$, $\times 10^{-8}$ м⁴	$\varphi''(\zeta)$	$\varphi'(\zeta)$	$\varphi(\xi)$	$F''(\zeta)$	$F'(\zeta)$	$F(\xi)$	$y(\zeta)$, $\times 10^{-7}$ м	$y(\zeta)/y(1)$
0	0.269	1	0	0	1	0	0	0	0
0.1	0.269	1	0.1	0.005	0.81	0.0905	0.0045	0.0196	0.0181
0.2	0.269	1	0.2	0.02	0.64	0.163	0.0172	0.0746	0.0689
0.3	0.269	1	0.3	0.045	0.49	0.2195	0.0363	0.1575	0.1454
0.4	0.269	1	0.4	0,08	0.36	0.262	0.0604	0.2618	0,2418
0.5	0.269	1	0.5	0.125	0.25	0.2925	0.0882	0.3820	0.3527
0.6	0.269	1	0.6	0.18	0.16	0.313	0.1184	0.5129	0.4735
0.7	0.269	1	0.7	0.245	0.09	0.3255	0.1505	0.6516	0.6017
0.8	0.269	1	0.8	0.32	0.04	0.332	0.1835	0.7942	0.7333
0.9	0.269	1	0.9	0.405	0.01	0.3345	0.2170	0.9387	0.8667
1	0.269	1	1	0.5	0	0.335	0.2505	1.0831	1

The average value of the intensity of the gas load along the length of the blade, in the circumferential direction

$$P_u = -\frac{G_r(C_{1u} - C_{2u})}{z(R_s - R_0)}.$$ (71)

$$P_a = \frac{\pi(R_s + R_0)}{z}(p_1 - p_2) + \frac{G_r(C_{1a} - C_{2a})}{z(R_s - R_0)}$$ (72)

The values of the bending moments M_{px} and M_{py} due to the action of dynamic loads of the steam P_u and P_a relative to the x and y axes:

$$M_{px} = -\frac{1}{2}P_u(R_s - R_i)^2,$$ (73)

$M_{p\eta} = M_{py}\cos\beta - M_{px}\sin\beta;$ (74)

$M_{p\xi} = M_{py}\sin\beta + M_{px}\cos\beta.$ (75)

where M_{jyi} - bending moment in the i-th section of the blade;

ΔM_1 - change in the moment of bending in the section i-1 from the action of the forces applied in the center of mass of the i-th section of the pen;

ΔM_2 - change in the bending moment in section i-1 from the acting force ΔP_j of the elementary mass located between sections i and i-1:

$$\Delta M_1 = -P_{ji}(x_i - x_{i-1}),$$

$$\Delta M_2 = -\frac{\rho \omega^2}{8}(F_i + F_{i-1})(R_i^2 - R_{i-1}^2)(x_i - x_{i-1}).$$

Having changed dependence (74, 75) taking into account the obtained values of ΔM_1 and ΔM_2, we write down the equation. The values obtained during its solution are summarized in Table 8. [104].

$$M_{jy(i-1)} = M_{jyi} - P_{jy}(x_i - x_{i-1}) - \frac{\rho \omega^2}{8}(F_i - F_{i-1})(R_i^2 - R_{i-1}^2)(x_i - x_{i-1}) \qquad (76)$$

Table 8. Calculated indicators for the sections of the blade

Geometric profile parameters	Blade section				
	2	3	4	5	6
M_{jyi} Hm	-36.3166	-20.035	-8.748	-2.157	0.001
$P_{ji} \cdot 10^{-3}$ H	119.697	86.381	55.801	27.057	0.000
$(x_i - x_{i-1}) \cdot 10^4$, m	1.595	1.595	1.595	1.595	1.595
ΔM_1, Hm	-19.085	-13.773	-8.898	-4.316	0.001
$(F_i + F_{i-1}) \cdot 10^4$ m^2	13.450	12.061	10.836	9.770	8,789
$R_i^2 - R_{i-1}^2$, m^2	0.049	0.052	0.055	0.058	0.061
$[6] \times [7] \times [4] \times 10^8$, m^5	1.039	0.995	0.948	0.999	0.861
$\Delta M = -\rho\omega 2/8 \times [8]$ Hm	-2.631	-2.511	-2.389	-2.279	-2.160
$M_{jy(i-1)}$, Hm	-58.028	-36.317	-20.035	-8.748	-2.157

Calculate the force from the center of mass acting in the plane R_{0y}

$$P_{jy(i-1)} = P_{jyi} + \Delta P_{jy} \ (i = 1\ldots2); \qquad (77)$$

where P_{jyi} is the component of the force P_{jy}, acting in a circle, applied in the center of mass of the i-th section of the blade feather;

ΔP_{jy} - component of the force ΔP_j of the elementary mass, located between sections i and i-1.

$$\Delta P_{jy} = \frac{\rho \omega^2}{4}(F_i - F_{i-1})(R_i - R_{i-1})(y_i + y_{i-1}). \qquad (78)$$

The calculated values by equality (78) are summarized in Table 9 [104].

Table 9. Calculation results

Geometric profile parameters	Blade section				
	2	**3**	**4**	**5**	**6**
P_{jyi},H	-44.617	-37.831	-27.661	-14.829	0.001
$(Fi+Fi-1)\cdot10^4m^2$	13.450	12.061	10.836	9.770	8.789
$(y_i+y_{i-1})\cdot10^4$,m	-0.940	-2.821	-4.702	-6.582	-8.463
$R_i - R_{i-1}$m	0.039	0.039	0.039	0.039	0.039
ΔP_{jy}.H	-2.525	-6.789	-10.171	-12.836	-14.831
$P_{jy(i-1)}$.H	-47.138	-44.616	-37.825	-27.658	-14.826

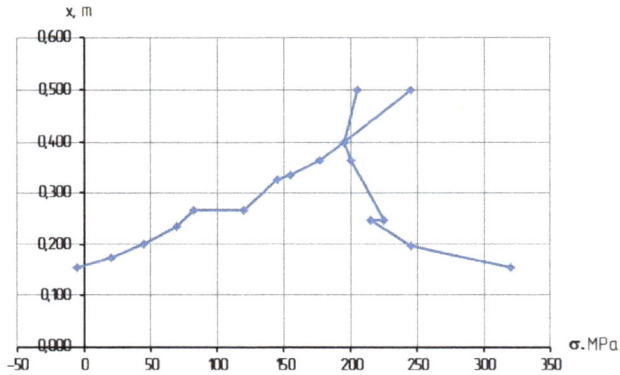

σ_p-- *radial stress across the disk,*

σ_r - *tangential stress across the disk*

Figure 14. Distribution of radial and tangential stress across the disk [103 p. 115]

The magnitude of the bending of the working blade connected by a shroud under the action of dynamic forces and their moments is adequately described by the equality:

$$\varphi'(\zeta)=\int_0^\zeta \varphi''(\zeta)\cdot d\zeta; \qquad F'(\zeta)=\int_0^\zeta F''(\zeta)\cdot d\zeta;$$

Assuming that the blade is of constant cross section, then $I(\zeta)=I_0$ at $0\leq\zeta\leq1$, hence $\varphi''(\zeta)=1$.

The power expended to overcome the forces of friction with the erosive-corrosive surface of the blade.

The power expended to overcome the forces of friction with the erosive-corrosive surface of the blade.

$$\Delta N_{\text{тв}} = (C(d-1)^2 + 3,44 \cdot 10^5 (1-\varepsilon - 0,5 \cdot \varepsilon_{\text{k}}) d(l_{21}^{1,5} + l_{22}^{1,5})) \cdot (\frac{u}{1000})^3 \cdot \frac{1}{\upsilon_{2t}} =$$
$$= (2750 \cdot (1,06 - 38,8 \cdot 10^{-3})^2 + 3,44 \cdot 10^5 (1 - 0,31 - 0,5 \cdot 0,62) \cdot 1,06 \cdot$$
$$\cdot (0,0291^{1,5} + 0,0484^{1,5})) \cdot (\frac{166,5}{1000})^3 (\frac{1}{0,1908}) = 121,91 \text{кВт}$$

2.4   Substantiation of the Nomenclature of Quality Indicators for Blades of Complex Geometry of Steam and Gas Turbines of CHP, Restored by a Highly Concentrated Source of Plasma Energy

One of the main indicators of any gas turbine plant is the resource and reliability [24 p. 402, 25 p. 327, 26 p. 532, 27 p. 277, 35 p. 29, 36 p. 670, 98 p. 25, 105 p. 31, 37 p. 20, 38 p. 130]. In many ways, these indicators are determined precisely by the rotor blades, since they are the most loaded element. During operation of the installation, the rotor blades experience cyclic, dynamic and static loads. In addition, working in an aggressive environment, the blades are highly susceptible to corrosion activity. Therefore, in the process of developing the design of CHP turbine blades, it is necessary to take into account the influence of the above factors at each stage of their life cycle [30 p. 23, 31 p. 15, 38 p. 72,106 p. 134]. Also, due to the constant change in temperature, large thermal stresses arise in the rotor blades [, 35 p. 29, 46 p. 73, 107 p. 767]. The temperature at which the blades are operated is 800-1250°C. Therefore, when operating at such high temperatures, the main deformations and fractures occur due to insufficient heat strength of the metal.

To date, the main material from which rotor blades are made is nickel-based alloys [23 p. 8, 24 p. 300, 28 p. 116, 34 p. 32, 35 p. 20, 39 p.2, 41 p. 42, 101 p. 112, 109 p. 64, 103 p. 11,]. Such alloys have a complex structure and phase composition. It is the structure that determines the main properties - fatigue resistance, ductility, heat strength.

The main structural elements of the blades are the tail section, the working section and the blade airfoil, which has a complex profile created by straight generatrices moving along a helical guide. When performing the plasma reduction process, it is necessary to comply with the requirements of regulatory documents for the shape of the turbine blade profile and observe the dimensional parameters. Since deviations from the requirements can lead to the development of material degradation processes and subsequent turbine failures [110].

For industrial products, including CHP turbine blades, a range of quality indicators is assigned that characterize the properties of the product to perform its functions and meet consumer requirements. An expedient choice of quality indicators that most fully characterize the object of control, in the future, allows for a comprehensive assessment of the quality of the product with high reliability.

Table 10 presents a selection of the main nomenclature of CHP turbine performance indicators. It is based on a complex indicator of reliability and a system of parameters characterizing the efficiency of turbine operation. These indicators are regulated by GOST 4.424 - 86 "Stationary steam turbines. Nomenclature of quality indicators.

Table 10. Typical nomenclature of quality indicators for CHP turbines

Parameter	Conditions	Formula	Notes
Main indicators of turbine reliability			
Reliability		$P(t) \approx \dfrac{N-m}{N}$	N is the number of observed products, m is the number of failed products.
Maintainability		$T_v = \dfrac{\sum_{i=1}^{m} \tau_i}{m}$	Assessment of the average recovery time
Availability factor	$0.85 \leq K_r \leq 1$	$K_r = \dfrac{T_0}{T_0 + T_v}$	T_o - mean time to failure, T_v - average recovery time of the product.
Technical utilization factor	$0.75 \leq K_r \leq 1$	$K_{TИ} = \dfrac{T_r}{T_r + T_{TO} + T_{rem}}$	T_r - time the product stays in working condition; T_{TO} - downtime due to maintenance; T_{rem} - repair time for the period of operation.
Operational			
Pressure (absolute) of fresh steam, MPa	18		
Steam saturation temperature, °C	555	$t_{pn} = t_p + \delta t_p$	Subheating to saturation temperature in a heater with a desuperheater is taken equal to $t_\delta = 2\ °C$ from the recommended range $t_\delta = 1...3\ °C$.
Rotor speed, rpm	3.000		
Turbine rated power, MW	110		
Live steam consumption at rated power, t/h	398		
Rated heating load, Gcal/h	175 (approximately 340)	$N_T = G_g L_{ad} h_T K_e$	G_g - Gas flow through the turbine; L_{ad} - Adiabatic work of steam expansion; K_e - pulse coefficient equal to 1.07 ... 1.4 - with pulse boost.

During operation at the stage of maintenance and repair of the turbine, these indicators determine: the average time to failure T_0; the average recovery time T_v in the performance of the product, etc. However, the presented parameters characterize the reliability of the turbine as a whole without taking into account the technical condition of its structural elements, such as the rotor and blades. The turbine produces energy by converting the pressure of the injected steam into the rotational movement of the rotor, the stator of which converts mechanical work into electrical and thermal energy. When interacting with the contact medium (saturated steam), the turbine blades provide a steady rotation of the rotor.

An analysis of the turbine operation indicators led to the conclusion that the pressure (absolute) of live steam, MPa, the rated power of the turbine, MW and the availability factor Kr directly depend on the technical condition of the turbine blades. However, GOST 4.424 - 86 and other regulatory documents (RD) do not provide for a deep technical assessment of the main elements of the turbine. Gaps in this direction do not allow to adequately assess the effective life of the turbine elements with minimal maintenance and operation costs. Thus, it is necessary to solve the scientific and technical problem of the lack of a unified system of criteria for assessing the technical condition of turbine blades, taking into account the depleted resource.

At present, there are no strict requirements for an objective assessment of the technical suitability of blades for further operation or repair and restoration work.

In industrial practice, the sorting of turbine blades is carried out on the basis of a comparison of indicators [60 p. 10, 111], presented in Table 11.

Table 11. Criteria for maintainability of turbine blades regulated by regulatory documents

Parameter	Conditions	Formula	Notes
Mechanical damage	Depth of notches, mm	0.5–1	
	Number of nicks per 1 mm	$n \leq 1$	
	equiaxed mechanical nicks per 1 mm, N_{ose}, pcs	$N_{ose} \leq 2$	
	- traces of grazing; - deformations of the output and input edges and the periphery of the working part	No	Determined visually by the operator-diagnostician
Metal washing	Washing depth, h_{pr} mm	$h_{pr} \leq 2$	
Corrosion ulcers	Ulcer diameter, mm	$\Delta_{ulc} \leq 0.5$	3 points and above
Erosive wear	Erosion spot diameter, mm	$\Delta_{ulc} \leq 0.5$	
Continuity of welded joints	No weld defect	The presence of continuity of welded joints	The presence of continuity of welded joints
Fatigue damage	Cracks		The crack length does not exceed 3 mm. The depth does not exceed 0.5 mm.

From the analysis of the main known indicators characterizing the maintainability of turbine blades, it has been established that all of them are indirect in nature and do not reflect the objectivity of the technical condition of the blades that have passed the operational period. For example, mechanical damage is random and has a wide range of manifestations. It is impossible to assess their real impact on the resource of work. Technical sources recommend assessing the depth of the risks, counting the number of nicks, etc. This stage of fault detection is implemented only visually without metrological support and is subjective. Moreover, erosive wear is determined by the diameter of the erosive spot within the operator's line of sight. A feature of the development of erosion and corrosion is the destruction of interatomic bonds due to oxidative processes and the chaotic distribution of foci in the structure of the material.

It is not visually possible to detect these defects and even more so to classify them in accordance with the criteria for assessing the technical condition of the blades. It was previously established that it is the structure of the metal of the blades that perceives dynamic and vibrational loads during the operation of the turbine. However, it is not possible to determine the progressive development of latent pitting or the initiation of fatigue microcracks without special equipment and a reasonably chosen technique. The proposed method of ultrasonic testing is very laborious and does not allow determining and interpreting the presence of fatigue defects in the structure of turbine blades.

Thus, it is necessary to develop a system of criteria for an adequate assessment of the technical condition of turbine blades, not only in terms of design parameters (shape, design geometry, linear dimensions and roughness), but also taking into account the features of fatigue changes in the structure of the blades. The complexity of developing criteria lies in the fact that it is necessary to develop a system for identifying and interpreting hidden defects with the possibility of generating recommendations for operation or restoration.

The system of proposed evaluation criteria should take into account the influence of the physical and mechanical properties of the material of the blades on their fatigue and vibration durability. Efficient operation of the turbine at rated power MW depends on the ability of the blades to perceive cyclically changing dynamic loads of the contact medium. This means that it is necessary to additionally investigate changes in the physical and mechanical properties of the blade that have passed a certain period of operation. To solve the problem, Table 12 proposes the main criteria for assessing the physical and mechanical properties of the blade material [112].

The proposed criteria for assessing the technical condition of the blades in conjunction with the use of standard parameters should fully characterize the fatigue.

For example, the accumulated metal fatigue can be determined based on the known physical phenomena of magnetic field scattering in the metal structure. As an assessment criterion, it is proposed to measure the magnetic field strength, A/m, followed by the determination of the fatigue limit, N/m^2 of the blade metal.

The peculiarity of the proposed criteria (Table 12) is that their evaluation will allow not only to identify and interpret hidden defects, but also to reasonably approach the choice of technological modes of the restoration process.

Table 12. Main criteria for selecting discarded turbine blades for subsequent restoration of their properties

Control Parameters	Symbol	Allowed Value	Method of Determination
Magnetic field strength, A/m	H_x	54 ± 1.42	Measuring
Blade material hardness, HRC	HRC	55-62	Measuring
Blade material microhardness, HV	HV	746-1.200	Measuring
Sectional area of corrosion-erosion-worn blade, mm^2	$\sum F_{total}$	$0.3\pm5\%$	$\sum F_{total} = F + 0,7d_p(\delta_1' + b_i'R_a)n_{plot}$ d_p – diameter, area with erosion-corrosion changes, mm; δ_1' - profile thickness, in the area with erosion-corrosion changes, mm; b_i'- blade profile depth in the area with erosion-corrosion changes, mm; R_a - surface roughness of the area with erosion-corrosion changes, microns.
Fatigue limit, N/m^2	σ_{-1}	$40\cdot10^6$	$\sigma_{-1} = 4.000 + 1/6\,\sigma_{max}$ σ_{max} – maximum stress in the area of action of stress concentrations.
Stress concentration factor	a_σ	1 ± 0.05	$a_\sigma = \dfrac{\sigma_{max}}{\sigma_H}$
Optimum length of a substandard blade suitable for recovery, m	L_{xv}^{kr}	$L_{xv}^{kr} \geq k_{pr}\cdot h_{kr}^*$	where L_{xv}^{kr} is the critical length of the feather from the tail of the blade required for restoration, m; k_{pr} - coefficient of suitability for recovery (according to experimental data), $k_{pr} = 0.969 \div 1.006$; h_{kr}^* - critical height of the blade airfoil tail, $h_{kr}^*=0.2...0.4L$, L - blade airfoil length, m.

Conclusion

1. Research has established that the resource life of turbines is determined by the reliability of the blade apparatus. Thus, the plasma reduction technology, including their subsequent control, must ensure that the quality parameters of the blades are consistently maintained.

2. It has been established that the blade airfoil has a complex shape of variable cross section, geometrically oriented in space with respect to the tail section, thus, the permissible deviations in the airfoil manufacturing are within 0.05-0.15 mm. The accuracy of manufacturing the blade tail with a lock is 0.01-0.032 mm.

3. According to the results of force and strength calculations of the CHP turbine blades, it was found that during operation under changing technological conditions, compressive and tensile stresses act on the blades. It has been established that the values of these stresses are

distributed unevenly along the entire length of the blade in different time intervals. This physical phenomenon, which is unstable in nature, leads to a sharp increase in dynamic loads in a short period of time, which reduces the efficiency of the turbine in transient acceleration and deceleration modes.

4. An improved mathematical model is proposed that describes changes in dynamic loads in different time intervals depending on operating modes and defects in the design geometry of the blades due to corrosion-erosion processes. The mathematical model made it possible to investigate the physical meaning of tensile stresses when the moments of inertia change. These indicators made it possible to reliably investigate the power of the turbine spent to overcome the resistance forces with worn blades during their operation. The established dependencies substantiated the need for an integrated approach in assessing the effective service life of turbine elements with minimal maintenance and operation costs.

5. An analysis of the turbine operation indicators led to the conclusion that the pressure (absolute) of live steam, MPa, the rated power of the turbine, MW and the availability factor Kr directly depend on the technical condition of the turbine blades. However, GOST 4.424 - 86 and other regulatory documents (ND) do not provide for a deep technical assessment of the main elements of the turbine. Thus, it is necessary to solve the scientific and technical problem of the lack of a unified system of criteria for assessing the technical condition of turbine blades, taking into account the depleted resource.

Chapter 3. Adaptation of the Technology of Recovery of the Phase Structure of Turbine Blades of CHP

3.1   Development of a Technological Process for Restoring the Phase Structure of Substandard Turbine Blades at CHP

The main problem in the technological process of reconditioning turbine blades of thermal power plants is the process of forming the optimal structural-phase state of the metal, which excludes the appearance of cold-drawn cracks and provides vibration resistance to dynamic loads [39 p. 2, 45 p. 80, 46 p. 49, 47 p. 124].

It has been established by research that residual stresses have a significant effect on the formation of cold cracks in welded structures made of medium and high alloy steels. Especially in products subjected to cyclic loads. At present, significant results have been achieved in the studies of laser-plasma recovery of working blades of turbines of thermal power plants. However, there are scientific and technological gaps regarding the influence of structural changes in the material of the blade airfoil on the formation of lingering cracks or mechanical tears and impact nicks [47 p. 130].

According to [111 p. 21] turbine blades are subjected to technical inspection during operation, according to the results of which they are sent for repair and restoration work or are rejected.

A well-known method of restoring Godovskaya G.V., Khafizova R.Kh. [39 p. 3] has problems in the use of dissimilar materials for surfacing (nickel-iron and cobalt-iron alloys, as well as nickel, cobalt and iron alloys), on the one hand, this technique provides a combination of ductility of the deposited ($\delta_5 = 15\%$) material and high hardness its surface (229 - 269 HRC), but at the same time leads to the formation of a heterophase layered structure, which reduces corrosion resistance [38 p. 120].

The previously described method for restoring the working blades of steam turbines proposed by Lappa V.A., Fedin I.V., Khromchenko F.A. [40 p. 15, 42 p. 25, 43 p. 10], is the need to remove the blades from the rotor and does not provide for heat treatment after surfacing, which complicates and increases the cost of work. Also, a significant disadvantage of this method is the high heterogeneity of the structural-phase composition and the high level of tensile residual stresses.

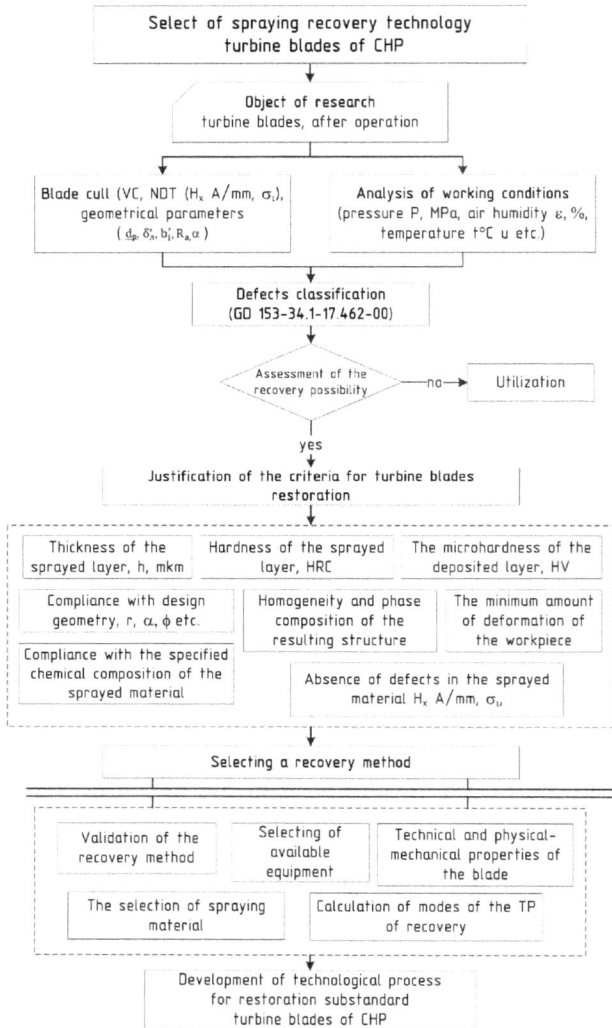

d_p is the diameter of the area with erosion-corrosion changes, mm; δ_l'- profile thickness, in the area with erosion-corrosion changes, mm; b_l'- depth of the blade profile in the area with erosion-corrosion changes, mm; R_a- surface roughness of the area with erosion-corrosion changes, microns

Figure 15. Block diagram of the algorithm for the justified choice of technology for the recovery of turbine blades by sputtering of CHP turbines [108]

Therefore, the task of developing an innovative technology that allows not only to restore the geometric parameters, but also to improve the quality of the structural and physical and mechanical properties of CHP turbine blades remains relevant and unresolved.

On the basis of an improved mathematical model of dynamic processes, the main indicators that affect the quality of turbine blade recovery are substantiated.

The restoration of worn parts is a complex organizational and technological process, in which, unlike the production of new parts, a formed but already worn part is used as a workpiece. The production experience of Remplazma LLP and the main results of the study on the renovation of blades made it possible to develop a structured model for the reasonable choice of technology for the recovery by spraying of substandard turbine blades of the CHP (Figure 15).

The defect detection of blades that have passed the operation cycle was carried out using the developed integrated assessment method, which makes it possible to identify stress concentration zones in the blade structure [102 p. 70, 113]. According to the research and the results of computer simulation in the SolidWorks environment, the main load falls from the middle of the blade airfoil to the upper end and leading edge (Figure 16).

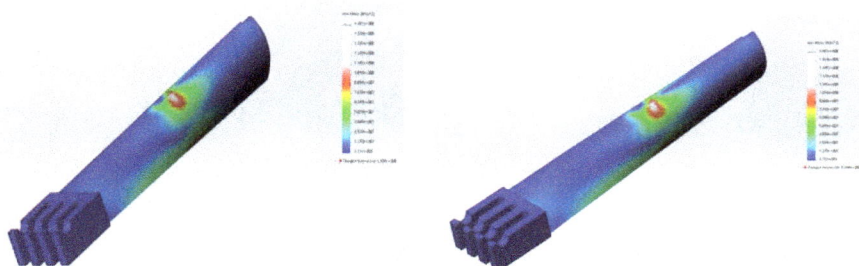

Figure 16. Stresses in the blade at a load of 75 kN

With an increase in the leverage of the load, tensile or compressive stresses increase. Also, the load is significantly increased by transient operation modes for a short period of time, which ultimately leads to a significant increase in internal bending stresses, displacements and deformations of the blade metal and to a fracture. Previously, blades with such damage were rejected. Based on the results of the studies and the calculations obtained, the technology for the restoration of turbine blades of CHPs by the "implantation" method was substantiated and developed under the following conditions [113 p. 62]:

$$L_{xv}^{kr} \geq k_{pr} \cdot h_{kr}^{*} \tag{79}$$

where L_{xv}^{kr} is the critical length of the feather from the tail of the blade required for restoration, m;

k_{pr} - coefficient of suitability for recovery (according to experimental data), $k_{pr} = 0.969$ $\div 1.006$;

h_{kr}^{*} - critical height of the blade airfoil tail, $h_{kr}^{*}=0.2...0.4L$, L - blade airfoil length, m.

3.2 Development of a Technological Process for Restoring the Phase Structure of Substandard Turbine Blades of CHP

Based on the results of fault detection, the criteria for blade restoration are determined, which allow choosing the most effective restoration technology. After analyzing the results of the studies, two possible recovery options were identified. The algorithm for choosing an acceptable recovery option is shown in Figure 17.

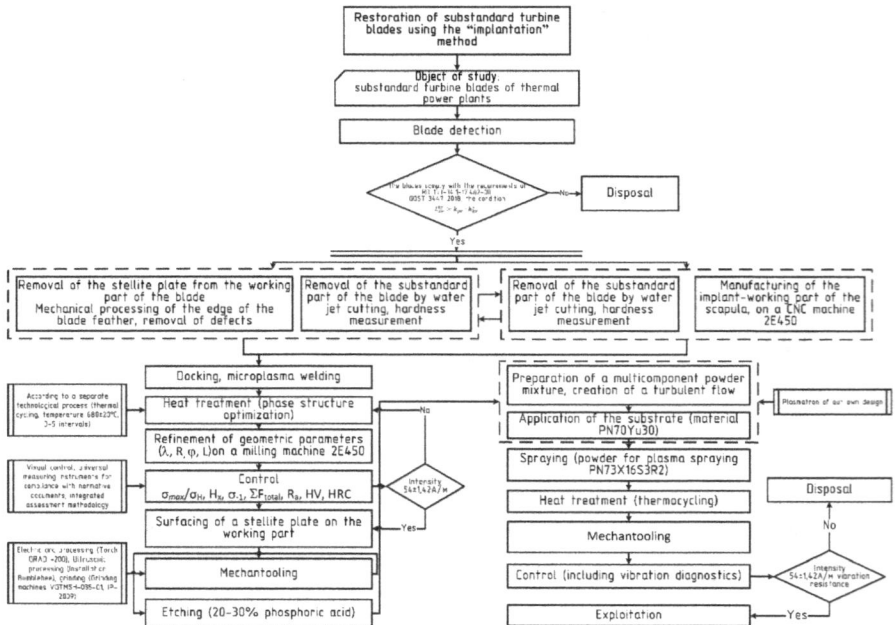

Figure 17. Algorithm for an innovative technological process for the restoration of substandard turbine blades of CHP

Restoration of the shoulder blades by the "Implantation" method. In this case, it is proposed to reconstruct the scapula using an implant. The essence of the process is as follows:

In the case when the blade airfoil has undergone unacceptable corrosion-erosion wear, it is removed, taking into account the design values (79) figure 18.

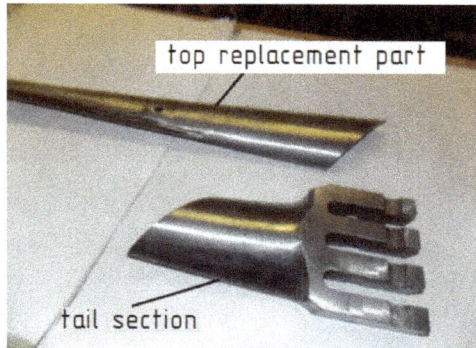

Figure 18. Cut off part of the blade and replaceable part

Further on a separate technological process, in accordance with the requirements of regulatory documents (Figure 19).

Based on practical experience, it has been established that it is not advisable to remove the defective part of the blade by mechanical or thermal means. Since when using these methods, the hardness of the edge of the blade increases by approximately 10% or more, compared with the hardness of the base metal. There is also a change in the structural-phase composition of the metal in the treated area, which in turn adversely affects the subsequent welding process.

Thus, to remove the substandard part of the blade, the waterjet cutting method was used on an APW 1525BA (WaterJet) installation. Installation power - 400 MPa, cutting speed 50 mm / min, nozzle diameter - 1 mm. The suspension speed is 1000 m/s, the power of the plant is 37 kW (Figure 20).

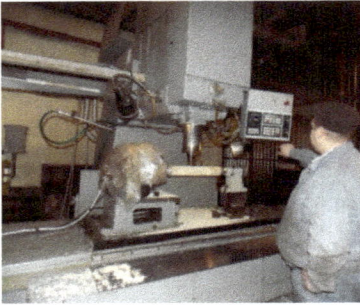

Figure 19. Making an implant   *Figure 20. Waterjet cutting*

During the subsequent measurement of hardness at the edge of the blade and the base metal (steel 15Kh11MF turbine T-100/120-130), the results HB=240-250 units were obtained, which is acceptable.

After manufacturing the missing element, it was adjusted in place, and after docking, microplasma welding was carried out at three points of the joint and then a weld was applied along the entire length of the joint (Figure 21).

Figure 21. Microplasma welding

The technological mode of welding was provided in such a way that the heat input did not exceed 20 cal/cm. The value of heat input Q was determined by the formula Q= $(125 \div 150)\tau$ cal/cm, where τ is the cross-sectional area of the weld (Figure 21).

After the completion of the previous stage, the heat treatment of the blade and the refinement of the geometric parameters of the blade were carried out in accordance with the requirements of the drawing.

Next, the edge of the blade airfoil was restored. In this case, defects were removed from the working part of the blade, including the stellite plate (if any). Next, the control of fatigue stresses in the metal structure was carried out using an integrated quality assessment method. In case of detection of zones of stress concentration, heat treatment of the blade was carried out. The stress state was equalized by thermal cycling:

a) the working blade (steel 15Kh11MF) was heated to a temperature of 680-700°C, followed by holding for 1-5 minutes per 1 mm of metal thickness;

b) cooling to a temperature above the temperature corresponding to the second limit of embrittlement;

c) the next heating up to 650-680°C and subsequent exposure according to the condition of the first heating and repeating steps a and b. The number of stages was taken equal to 3-5 heating and cooling cycles.

After leveling and reaching the allowable residual stresses, the surface of the blade was treated with an electric arc (plasma) with transverse oscillations. At the same time, metal penetration was ensured to a depth of 0.1-0.2 of the thickness of the layer planned to be welded. The transverse oscillations of the arc with an amplitude were 2-4 diameters of the non-consumable electrode.

After mechanical treatment, the area of the blade with the deposited plate was subjected to ultrasonic impact treatment. Next, the surface was polished until the roughness parameter was not lower than $R_z = 20$ μm. The blade was then etched and a nickel base alloy sublayer was implanted. Next, sputtering was carried out with concomitant reflow. After the operations, the blade was subjected to heat treatment in order to reduce the level of residual stresses. This was followed by mechanical processing and control.

Upon completion of the heat treatment, control tests of the restored blade were carried out. The developed integrated evaluation method was used to evaluate internal stresses and evaluate defects. Also, universal measuring instruments check the correspondence of the geometrical parameters of the blade (length, bend angle, roughness, etc.).

If there are no defects, a substrate of PN70Yu30 material is applied to the blade, while the blade is heated to 150-200°C and then a protective layer of powder material of the PN73Kh16S3R2 brand is sprayed.

After spraying and melting of the protective layer, the heat treatment of the blade was carried out in the appropriate modes. At the end of the cycle, the blades were subjected to mechanical processing (grinding, polishing) for compliance with the regulatory documentation and the standard.

Next, non-destructive testing was carried out by the method of metal magnetic memory and eddy current testing, and the hardness of the restored surface of the blade edges and in the heat-affected zone was also studied. After a comprehensive check of the repair quality, the blade went through vibration tests (Figure 22).

Figure 22. Vibration testing of blades on the bench

The blades passed the tests in accordance with the requirements of regulatory documents [114-116].

During the experiment at the Remplasma LLP enterprise, it was found that the main parameters of the plasmatron operation are the spraying distance, the particle speed and the granulometric composition of the composition, by changing which it is possible to increase the powder utilization factor, adhesion strength and coating hardness (Figures 23, 24) [113 p. 63].

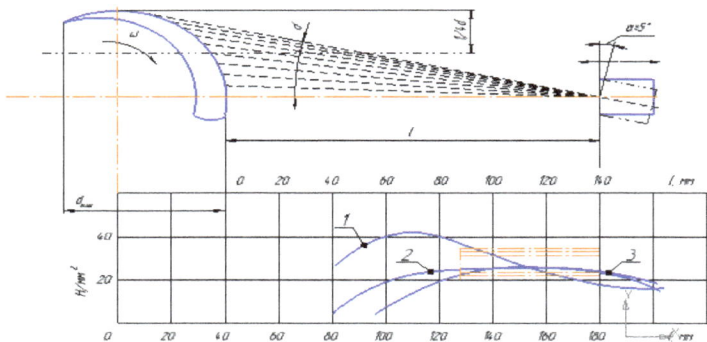

1 - at a speed of 55 rpm; 2 - at a speed of 30 rpm, 3 - at a speed of 7 rpm

Figure 23. Dependence of coating adhesion strength on spraying distance l mm

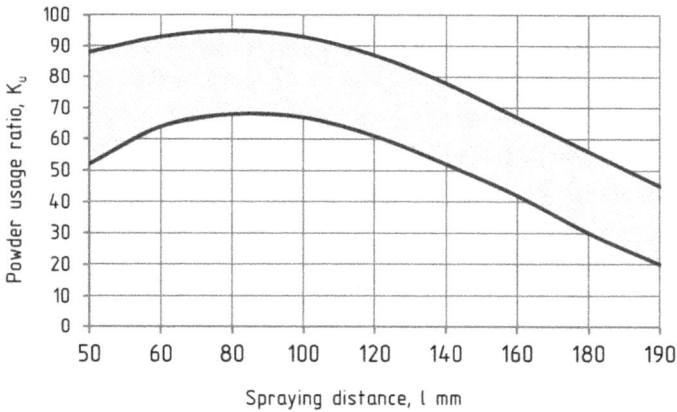

Figure 24. The dependence of the coefficient of use of powder PN55T45

Particles that are too large in size do not have time to heat up, as a result of which they do not form a strong bond with the sprayed surface, or the particles bounce when they collide with it. Small particles, in turn, do not have time to acquire a sufficient level of kinetic energy necessary for the formation of a strong bond with the sprayed surface. Moreover, small particles have time to cool before they reach the surface of the part. Typically, spraying powders with a particle size of 25 to 250 microns are used. Thus, in order to obtain a coating that is uniform in structure and thickness, it is necessary to carry out deposition at the smallest possible distance from the melting point to the surface to be restored. At the same time, an excessively small distance will cause overheating of the restored part and the appearance of pores, and as a result, cracks in the sprayed layer.

According to the results of calculations, it was found that the optimal distance is 80-120 mm (Figure 23), but depending on the characteristics of the technical characteristics of the restored object, it can be increased [47 p. 220, 117 p. 139].

Studies have established that the highest values of the hardness of the sprayed layer were achieved by adding propane as a plasma gas (POG). As practice has shown, with an increase in the amount of propane, the porosity of the coating decreases. At the same time, a high concentration of propane in the POG composition leads to the appearance of cracks on the surface of the resulting coating, in connection with this, when restoring the working blades of CHP turbines, the propane consumption Q must be strictly established for each brand of material (Figure 25). In connection with the physical phenomena occurring during deposition, with an increase in the thickness of the coating layer, the likelihood of the formation of SCZs increases, which in turn lead to a deterioration in the adhesive properties of the applied coating with the substrate.

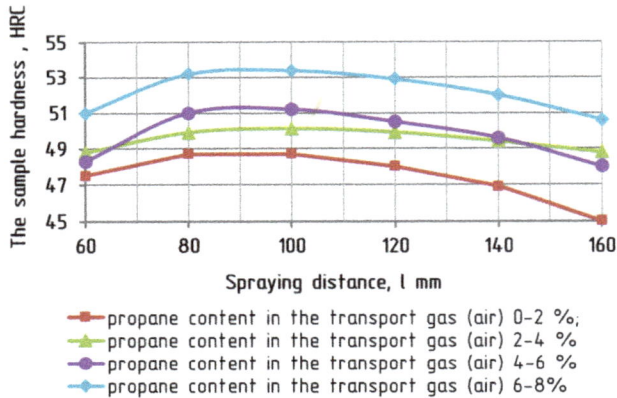

1 - propane content in the plasma gas (argon) 0-2%; 2 - propane content in the plasma gas (argon) 2-4%; 3 - propane content in the plasma gas (argon) 4-6%; 4 - propane content in the plasma gas (argon) 6-8%

Figure 25. Dependence of the hardness of the coating from powder PN55T45 on the distance of spraying

It has been experimentally proved that the thickness of the coating should not exceed 1 mm, since the action of dynamic loads falls on the base material of the part. The coating gives the part the necessary properties [40 p.3, 42 p. 10, 43 p. 15, 44 p. 012037-4].

Based on the results of metallographic studies, the optimal phase structure of the base of the repaired part was determined, which ensures a high level of reliability under the action of cyclically changing dynamic loads. To achieve high physical and mechanical properties of the restored blade structure as close as possible to the original ones, a technology has been developed for restoring substandard CHP turbine blades, which involves the introduction of an implant followed by thermal cycling.

In the course of the research, the dependences of the influence of technological regimes on the quality of recovery were established, which make it possible to reliably determine the optimal values of recovery modes.

3.3   Development of the Plasmatron Design and Calculation of its Parameters

The efficiency of the plasmatron is one of the most important principles of its design. Achievement in the process of designing the planned indicators of functionality implies compliance with the criteria of performance, reliability, quality and safety. The solution of this problem is often associated with the need to apply optimization design methods, since the efficiency criteria, as a rule, are contradictory in relation to the cost and safety criteria [118].

A number of parameters that determine the quality and efficiency of the plasma process are regulated by standards. In welding technologies, one should be guided by GOST 4.140-85 "System of product quality indicators. Electric welding equipment. Nomenclature of indicators", which defines 7 groups as the main indicators: 1) appointments; 2) reliability; 3) economical use of raw materials, materials, fuel, energy and labor resources; 4) manufacturability; 5) standardization and unification; 6) patent and legal; 7) security [118 p. 150].

The main parameters according to this standard are: the highest speed, specific energy consumption and efficiency, some of the criteria are given in Figure 26 as efficiency criteria specific to plasma processes. It is obvious that a number of stated criteria (energy consumption or current-voltage characteristics (CVC), efficiency; plasma jet parameters that determine the speed, thickness and width of the deposited material) can be evaluated in the process of designing plasma equipment.

The functional parameters of the plasmatron, which determine its performance, are associated with its design features and technological characteristics, as well as with the properties of the metal being processed (grade, brand, thickness, etc.). The experimental material accumulated to date makes it possible to set the optimal technological regimes of plasma action on the material, taking into account design changes (as a rule, the diameter of the plasmatron nozzle).

The book solves the problem of developing equipment for restoring the physical and mechanical properties and design geometry of the blades of the 27th stage of the T-100/120-130 turbine, steel grade 15X11MF, by highly concentrated sources of plasma energy.

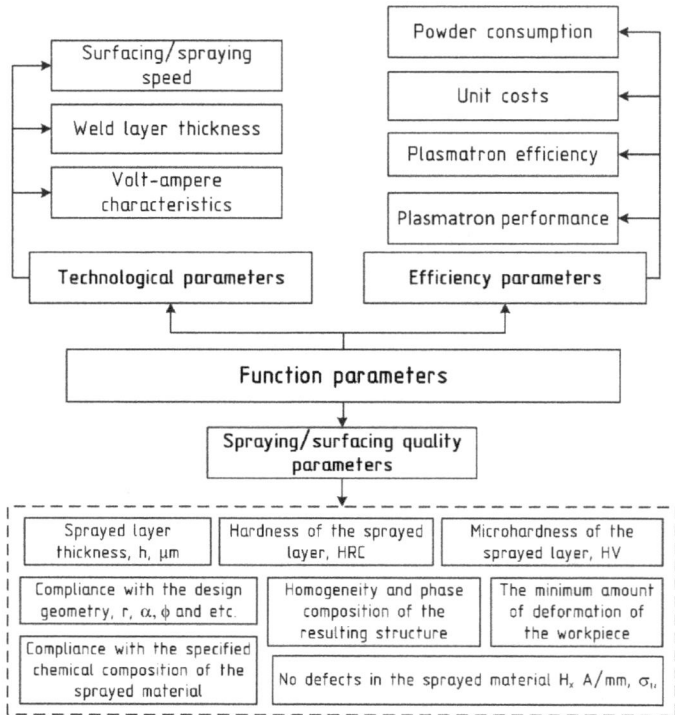

Figure 26. Plasma surfacing/spraying efficiency parameters

One of the main components of the plasmatron is the nozzle assembly, when designing it, it is necessary to take into account gas-dynamic, electrical and thermal factors of arc formation. The technique for calculating the nozzle assembly was created at the end of the last century and is currently the basis for the design of most modern plasmatrons. However, its application currently requires correction, taking into account new design solutions and research methods.

Among modern research methods, works on numerical modeling of processes in the nozzle assembly of plasmatrones predominate [118 p. 153, 119-122]

The process of developing a design model for the plasmatron began with the calculation of technological and design parameters. Knowledge of the geometric characteristics of the plasmatron model increases the potential for prototyping the design and technological complex for the plasma recovery of turbine blades using an effective virtual simulation technique. Therefore, one of the primary tasks is to calculate the main parameters of a portable plasma generator.

It is planned to use nickel-based refractory powders, tungsten carbides, solders as materials for restoration (materials used in restoration: powder for plasma spraying PG-10N-01; powder PN73X16S3R2; powder UTP HA - 6, Ni-Cr-B-Si -Fe; tungsten powder; Pr35 solder).

It is planned to use a direct current source with an open-circuit voltage of at least 300 V and an installed power of at least 80 kW as a power source for the plasmatron.

Nominal values of climatic factors: air temperature: lower value +1 °C; upper value +35°C; the upper value of air humidity is 80% at a temperature of +25°C.

The developed portable plasmatron, according to the set functional and technical tasks, should have the main technical characteristics - plasma temperature, T and plasma output pressure, P. In this case, the average temperature of the plasma jet cross section, T_{av}, is also taken into account in the calculations, it is also necessary to control the heat balance of the plasma substance. For this, it is important to take into account the maximum possible plasma temperature T_{max} (in some cases, the temperature of the plasma jet axis is used for calculations). As a result, thermodynamic equilibrium is provided at the output of the plasmatron, so these parameters (temperature and pressure) can fully characterize the composition of the plasma jet. In this case, the plasma parameters are tabular values [120 p. 206]. To improve the efficiency of the plasma formation process, it is necessary to take into account the type and characteristics of the gas used and the degree of its ionization, $\alpha < 1$.

The calculation of the plasmatron parameters and the overall dimensions of the discharge channel is presented below:

- maximum speed of sound at 5000 K $v_{cr} = 1447.1$ m/s;
- maximum air density at 5000 K $\rho_{cr} = 0.0575$ kg/m³;
- initial enthalpy $i_{in} = 3 \cdot 10^5$ J/kg;
- final enthalpy $i_v = 10.252 \cdot 10^6$ J/kg.

To calculate the electrical and thermal characteristics of the plasmatron, we use the following system of equations.

For a linear circuit plasmatron with a self-adjusting average arc length, the equation for the static current-voltage characteristic of an air-stabilized arc with direct polarity can be written as:

$$U = 3.060 \cdot \left(\frac{I^2}{G \cdot d} \right)^{-0,17} \cdot \left(\frac{G}{d} \right)^{0,12} \cdot (P \cdot d)^{0,5} ; \tag{80}$$

where I is the current strength, A

G – gas consumption, kg/s

d is the discharge channel diameter, m

P is the air pressure at the outlet of the plasmatron, Pa

The thermal efficiency of the plasmatron is determined by the following formula:

$$\frac{1-\eta}{\eta} = 5.85 \cdot 10^{-5} \cdot \left(\frac{I^2}{G \cdot d} \right)^{0,27} \cdot \left(\frac{G}{d} \right)^{-0,27} \cdot (P \cdot d)^{0,30} \cdot \left(\frac{l}{d} \right)^{0,50} ; \tag{81}$$

where G is gas consumption, kg/s

l/d – relative length of the output electrode, m

P is the air pressure at the outlet of the plasmatron, Pa

The energy of the flowing jet is calculated by the formula:

$$U \cdot I \cdot \eta = G \cdot (i_{_6} - i_{_{6x}}); \tag{82}$$

The power invested in the arc is determined by the formula:

$$N_q = U \cdot I. \tag{83}$$

The resulting system of equalities is not closed, therefore, we introduce two additional conditions that will establish a connection between the indicators being determined. According to experimental data, at the temperature of the plasma jet T = (3000 - 4000)K and pressure P = $1 \div 5) \cdot 10^5$ Pa, the length of the output electrode is determined: $\bar{l} = l / d = 20$ mm. The following condition characterizes the absence of thermal blocking in the channel of the cylindrical electrode

Therefore, the value of the internal diameter of the electrode is taken to be 10-30% more than the maximum allowable. Thus, the optimal electrode diameter will be d = $1.3 d_{cr}$

Taking into account the obtained data, the value of the diameter of the discharge channel of the plasmatron can be calculated using the following formula:

$$d = 1.3 \cdot 2 \sqrt{\frac{G}{\pi \cdot \rho_{cr} \cdot v_{cr}}}, \tag{84}$$

substituting the original values, we get:

$$d = 2 \cdot 1.3 \sqrt{\frac{2 \cdot 10^{-3}}{3.14 \cdot 0.575 \cdot 1,4471}} = 7.2 \cdot 10^{-3} \, м \tag{85}$$

The compiled system of equations was solved using the MathCAD software product. According to the calculation results, the following parameter values were obtained: arc voltage, U, was 517 V; current I = 57 A; thermal efficiency η = 68%; discharge channel length l = 0.14 m; plasmatron power N_q = U·I = 517 · 57 = 29.4 kW.

Let us conditionally divide the length of the discharge channel into two components: the length of the cathode l_k = 0.07 m and the length of the anode l_a = 0.07 m.

One of the priority tasks is to stabilize the flow of the carrier gas with powder material and to exclude erosion processes in the anode and cathode units of the plasmatron. This problem is solved by the original design of a three-way mixer. The proposed device operates on the principle of ensuring the turbulence of the flow due to its passage through technological openings that provide swirling of vortex flows.

The efficiency of the process is realized while maintaining the optimal value of the air velocity v_ϕ= 150-200 m/s, at the exit from the holes of the swirl ring. It has been established that the optimal value of the air velocity regime is v_ϕ= 175 m/s. Since the gas passes through the sections of three holes, to calculate the diameter of the holes we take the gas flow rate equal to:

$$d_\phi = \sqrt{\frac{3 \cdot G}{n \cdot \rho \cdot v_\phi \cdot \pi}}, \tag{86}$$

where n is the number of holes in the mixer, equal to three;

ρ - air density under normal conditions, ρ= 1.29 kg/m³

Thus, we get:

$$d_\phi = \sqrt{\frac{3 \cdot 2 \cdot 10^{-3}}{3 \cdot 1.29 \cdot 175 \cdot 3.14}} = 3.2 \cdot 10^{-3} \text{ м}$$

Argon is used as a plasma-forming gas, and air is used as a transport gas.

For efficient operation of the plasmatron, it is necessary to implement the following processes:

1) creation of an electric discharge with the required parameters;

2) the formation of a plasma flow as a result of the interaction of the plasma-forming gas and the discharge of a tungsten rod;

3) protection of the plasmatron body from overheating by the plasma jet;

4) temperature regime of electrodes, distribution of discharges on them, protection against oxidative processes. These functionalities are provided by the following subsystems:

a) power supply subsystem. Contains electrodes and auxiliary structural elements for ignition of the discharge;

b) systems of transporting and plasma-forming gases;

c) a cooling system, including cooling jackets for heat-stressed structural elements, channels for organizing the flow of a cooling medium, seals, and connecting elements to an external heat removal system.

These elements are implemented as a technological structure of the plasmatron design. In order to ensure the manufacturability of the assembly process, the design of the plasmatron in these nodes place all the functional elements of the systems. General requirements for the flame torch are presented in Table 13.

High-temperature arc "cold" plasma (up to 30,000°C) is used as a heating source in the process of plasma surfacing and spraying.

In the course of experiments on the restoration of substandard turbine blades, it was decided to modernize this model of the plasmatron by introducing a three-way vortex mixer into the gas-air system, which makes it possible to control the modes of the restoration process.

In order to increase the energy efficiency of the technological process for the restoration of turbine blades at the CHP, it is proposed to modernize the plasmatron by developing a three-way vortex mixer, which makes it possible to use multicomponent mixtures of powders. The proposed design will make it possible to simulate the variability of the properties of the resulting surface layer during the deposition process and thereby improve the necessary design and technological parameters of the parts to be restored.

Table 13. General hardware requirements

Name	Meaning
Operating current, A	50-200
Operating voltage on the plasmatron (when working in air), V	270-550
Plasmatron power, kW	30
Operating mode, PV,%	100
Plasma gas	Argon
Applicable	Nitrogen
Carrier gas	Air
Plasma gas consumption, g/s	0.8-3
Shielding gas consumption, g/s	0.15-0.45
Conveying gas consumption, g/s	0.3-0.9
Water consumption, g/s	270
Productivity for aluminum oxide, kg/h	10
Productivity for metal powders	30
Powder utilization rate	0.7
Plasmatron mass, kg	3

Structural changes in the technological elements of the plasmatron significantly change its structural characteristics, which leads to a change in the gas-dynamic processes occurring in it. The changes result in the appearance of additional hydraulic and gas resistance to the flow of the carrier gas with the powder mixture. In order to ensure optimal technological regimes, one of the primary tasks in the modernization of the plasmatron is to carry out a gas-dynamic calculation of each element of the gas-air path. Having studied the principles of pressure distribution, flow rate and change in the flow rate of the CG and powder composition, it will be possible to design the optimal design of the plasmatron without compromising the standard technical characteristics.

3.4 Researching of the Gas-Dynamic Features of the Structural Elements of the Gas-Air Path of the Plasmatron

The formation of a carrier gas flow (CG) with powder material is influenced by the processes associated with the gas flow rate in various parts of the gas-air path (GAP) of the device. These features include the expansion and contraction of the CG flow in different areas, changes in the turbulence of the flow and the direction of its flow. Thus, the patterns of change in the process of formation of the CG flow depend on the geometric dimensions of the sections of the gas-air path of the plasmatron, the properties of the gas used, the powder material, and their parameters [123].

Gas-air paths of plasmatrons, which have received the widest distribution, as indicated in [118 p. 160, 124] consist of sections of variable cross section, which leads to the formation of turbulent flows, pressure drops in sections with different cross sections, as well as a step change in the velocity of the transport gas and, as a result, plasma microexplosions and acoustic emission ejection of the transport gas from the nozzle.

Changes in the design of the GAP due to the modernization of the plasmatron should not lead to cardinal changes in the flow and loss of pressure at the outlet of the plasmatron, which in turn can reduce its efficiency. Thus, it is necessary to analyze the gas flow in the modernized design of the plasmatron. In the course of the analysis, a number of parameters were studied, the main of which are the speed and pressure of the transport gas flow, the uniformity of the distribution of the air-powder mixture over the cross section of the plasmatron channels, and the smoothness of the jet. The listed parameters strongly depend on the design features of the plasmatron.

The GAT design diagram of the plasmatron is a complex system of channels that provide the supply of powder material for plasma spraying. In which six sections can be conditionally distinguished: 1 - inlet (transporting line), 2 - three-way vortex mixer, 3 - transportation channel, 4 - swirler, 5 - nozzle, 6 - outlet (Figure 27).

Each section of the GAT plasmatron has individual geometric parameters that characterize its shape and dimensions, on which the cross-sectional area of each channel depends. Figure 1 shows the dependence of the change in gas velocity and pressure on the cross-sectional areas of each section in the direction from inlet to outlet (from left to right).

The considered sections of the gas-air path of the plasmatron are characterized by an original design, which, in turn, regulates the parameters of the gas flow in this section.

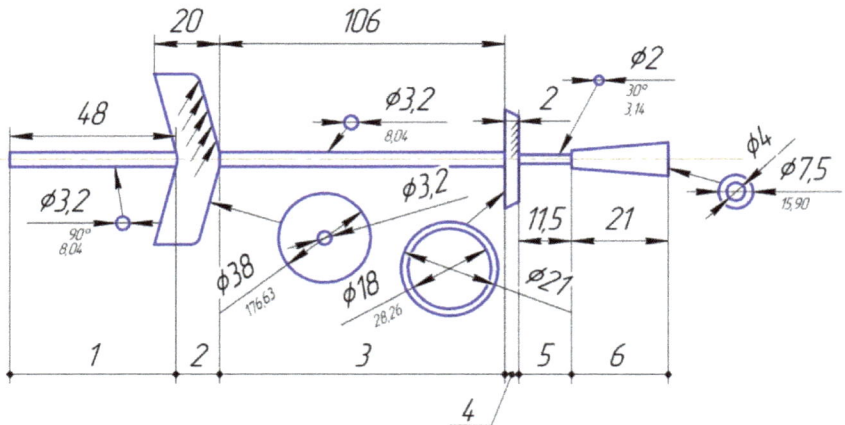

Figure 27. Structural scheme of the GAT plasmatron

The first section of the GAT of the plasmatron, in the presented system, is characterized by a constant cross-sectional area, through which the CG and powder are fed.

The main purpose of the second section, where the proposed mixer is located, is to ensure flow turbulence, uniform mixing of powder materials and its direction into the transportation channel.

The third section is an element of the gas-vortex system - a transportation channel designed to equalize the flow of the gas-powder mixture.

The fourth section is designed to form a swirling flow of the gas-powder mixture and feed it into the nozzle. In this section, the flow velocity increases sharply and a pressure drop occurs due to changes in the cross-sectional area.

In the fifth section, a jet of gas and powder composition under pressure exits the plasmatron nozzle.

In the sixth section, the gas-powder mixture enters the plasma-forming gas flow, which leads to the formation of a plasma jet saturated with the elements of the powder composition.

The regularities of the flow of the CG and the gas-powder mixture and its dynamic model are described by the Navier-Stokes dependences. Equations used to characterize the properties of the gas flow and reference data on thermal conductivity, viscosity and temperature are also applied when modeling a stream jet. To describe the change in the nature of the flow, its transition from a laminar state to a turbulent one, the allowed values of the Reynolds numbers are used.

Defined [118 p. 161] that it is possible to evaluate the efficiency of plasmatron design using theoretical dependencies and ratios of known criteria [118 p. 162]. These dependences make it possible to reliably determine the technological and efficiency parameters of the plasmatron. Due to the design features of the transporting line, it is necessary to carry out a gas-dynamic calculation of the elements of the gas-air path.

Since the carrier gas (CG) practically does not change its temperature when passing through the GAP of the plasmatron, except for the region of interaction with the plasma arc in the nozzle assembly, the calculations were performed on cold gas (T = 300 K) at a constant value of dynamic viscosity. As a carrier gas, a mixture of air and propane is used, the viscosity of which was averaged over the studied temperature and pressure range ($0.3 \div 0.5$ MPa). Since the mean free path of air molecules at the indicated pressures is from 0.01 to 0.1 mm, which is much less than the characteristic hydraulic diameters of the GAT ($2 \div 10$ mm), the gas in the plasmatron is not Knudsen and its viscosity strongly depends on pressure (($v \sim$ P-1) and weaker - on temperature ($v \sim$ T1/2) [117 p. 327]. Taking into account these dependences, the values of the coefficients of dynamic $\mu = 2 \cdot 01^{-5}$ Pa·s and kinematic viscosity $\nu = 5 \cdot 10^{-6}$ m²/s, respectively, were chosen in the calculations.

The gas-air path of such a plasmatron is a channel of changing shape with sharp changes in the cross-sectional area (Figure 27). When calculating the hydrodynamic parameters, the plasmatron path was divided into a number of transitions with constant or smoothly changing characteristics. A preliminary analysis made according to the continuity equation (constant mass flow rate) at the operating value of the plasma gas mass flow rate in the plasmatron $G = 1.2 \cdot 10\text{-}3$ kg/s showed that the gas flow is turbulent with Reynolds numbers (Re) from 1,000 to 320,000, and cross-sectional average velocities from 10 to 300 m/s. The calculation of gas-dynamic parameters in the turbulent flow regime was carried out taking into account gas-dynamic functions, according to.

For each section of the gas turbine, dynamic head losses were calculated, including inertial losses due to local hydraulic resistances and linear pressure losses due to viscosity, the cross-sectional average axial gas velocities and the nature of the flow were determined by the Reynolds number.

The calculation of hydraulic losses for each transition was carried out according to the formulas:

$$\Delta P_z = \sum_i \frac{\Delta P_i}{\Delta x_i} \cdot L_i + \sum_j \Delta P_j$$

where $\Delta P_j = \xi \times \dfrac{\rho_i v^2}{2}$ - local resistances taking into account inertial losses,

ξ - resistance coefficient depending on the shape of the hydraulic transition; Li – characteristic transition length;

$\dfrac{\Delta P_i}{\Delta x_i} = \zeta \cdot \dfrac{\rho_i v^2}{2 D_r}$ - linear pressure loss due to viscosity,

$\zeta = \dfrac{\Delta}{\text{Re}}$ - with laminar flow,

$\zeta = f\left(\text{Re}, \dfrac{\Delta}{D_r}\right)$ - in turbulent flow, Δ - ruggedness, $\text{Re} = \dfrac{v D_r}{\nu} = \dfrac{\rho v D_r}{\eta}$ - Reynolds number,

$D_r = \dfrac{4 F_s}{P_s}$ - characteristic hydraulic diameter of the transition and the corresponding gas path, introduced to assess the effect of the section shape on the loss of hydraulic head;

Fs - cross-sectional area, Ps - section perimeter;

$\rho_i = \rho_0 \left(\dfrac{P_i}{P_0}\right)^{\frac{1}{\gamma}}$ - density in the i-th section, Pi and P$_0$- pressure at the i-th section and at the entrance to the plasmatron, $\gamma \approx 1,4$ – gas adiabatic index.

The gas-dynamic parameters of the gas in separate sections were calculated sequentially, starting from the entrance to the plasmatron, corrected for the pressure loss in the previous sections of the gas turbine. The dependence of viscous losses on the Reynolds number characteristic of the selected profile was taken into account. Since the nature of the CG flow of the developed plasmatron is a turbulent flow, the resistance parameters ξ practically do not depend on the properties of the gas and its velocity. The calculation of the parameters of the change in the nature of the flow of the CG flow depending on the change in the cross-sectional area of each section was carried out according to the average indicators of the change in their sections in places of sharp drops. In areas with smooth drops, the nature of the change in the hydraulic diameter Dg was used to evaluate them [118 p. 211]. The hydraulic parameters of each transition were calculated sequentially, starting from the entrance to the plasmatron, where the pressure P0 was experimentally controlled.

For calculations and comparative analysis, as well as for numerical modeling, the average values of the gas-dynamic parameters of the gas were taken, which are acceptable for most of the investigated plasmatrons G=0.0005 kg/s.

The results of the gas-dynamic calculation are obtained based on the design in Figure 27 and are shown in Table 14.

In order to improve the efficiency of the plasmatron, the need to change some of its structural elements has been proved by calculation. During the modernization of the plasmatron, the principles of the integrity of its structure were observed, which ensures the preservation of the optimal parameters of the technological process. Studies have established that the quality of the technological recovery process is affected by regime parameters, the main of which depend on the design of the plasmatron. For example, high values of adhesion are achieved by high-speed regimes of the plasma jet. The speed of the plasma jet and its thermodynamic characteristics depend on the type and cross-sectional area of technological openings of the gas-air path.

Analyzing the results presented in Table 1, it was found that the gas-dynamic parameters of the carrier gas and the powder mixture in certain sections of the GWP of the plasmatron change in proportion to the pressure (Pin = 0.3÷0.5 MPa). At Pin = 0.3 MPa, the flow velocity of the CG in the conveying line is 153 m/s with a cross-sectional area F = 8.04 mm² when it enters a three-way vortex mixer, it decreases to 7 m/s, which is associated with the appearance of resistance forces that increase losses speed mode. This design feature is inevitable, since there is a technological need for a uniform distribution of the fractional composition of the powder composition.

Table 14 - Gas-dynamic parameters of the carrier gas and powder mixture in separate sections of the GAP of the plasmatron

Options		Lot number						Total values of indicators
		1	2	3	4	5	6	
		inlet (transporting line)	three-way vortex mixer	transportation channel	swirler	nozzle	outlet	
Cross-sectional area, F mm2		8.04	176.63	8.04	28.26	3.14	15,90	-
Hole diameter, D mm		3.2	15	3,2	6	2	4.5	-
Section length, L mm		48	20	106	2	11.5	21	208.5
Gas flow rate, V m/s	Pin=0.3	153	7	185	56	440	255	-
	Pin=0.5	135	6	151	45	120	121	-
Re·103	Pin=0.3	85	4	85	97	154	55	
	Pin=0.5	108	7	108	122	108	67	-
Pressure difference P MPa	Pin=0.3	31.1	0.3	32.4	17	138.6	21	240.4
	Pin=0.5	34.6	0.3	36.3	21	145.2	18	255.4

To ensure efficient technological conditions, a transportation channel with a cross-sectional area of 8.04 mm2 is structurally provided, which makes it possible to provide a flow rate of up to 185 m/s in a short period of time. To ensure high values of adhesion, the effect of "implantation of a molten particle" is used, which is created in the swirler (section 4). Vortex mixing of the powder composition inevitably leads to a turbulent flow at Re = 97·103, which creates additional resistance forces and leads to a decrease in pressure and gas flow velocity to 56 m/s. At this stage, one of the priority tasks is not only to improve the physical and mechanical properties of the coating, but also to effectively control the modes of the technological process. After the loss of speed in section 4, it is necessary to compensate for this parameter, for which the design of the working nozzle provides additional inlet sections that increase the speed of the gas mixture to 440 m/s. Next, the gas-powder mixture enters the plasma-forming gas flow, which leads to the formation of a plasma jet saturated with the components of the powder mixture. The shape of the nozzle outlet channel also allows you to adjust the speed of the plasma flow within 350 m/s.

The data obtained from the analysis of the design features of the GAT of the modernized plasmatron prove that the use of a device for mixing powder materials in the plasmatron provides the ability to control the optimal technological modes that affect the efficiency of its operation.

The proposed design makes it possible to control the deposition process by adjusting the proportions of a multicomponent powder material, which makes it possible to simulate the necessary physical and mechanical properties of the resulting structure and surface of the part.

The design of a three-way eddy current mixer is a conical cup with three holes located at an angle of 30° for the introduction of powder materials with a carrier gas, Figure 28 [125-129].

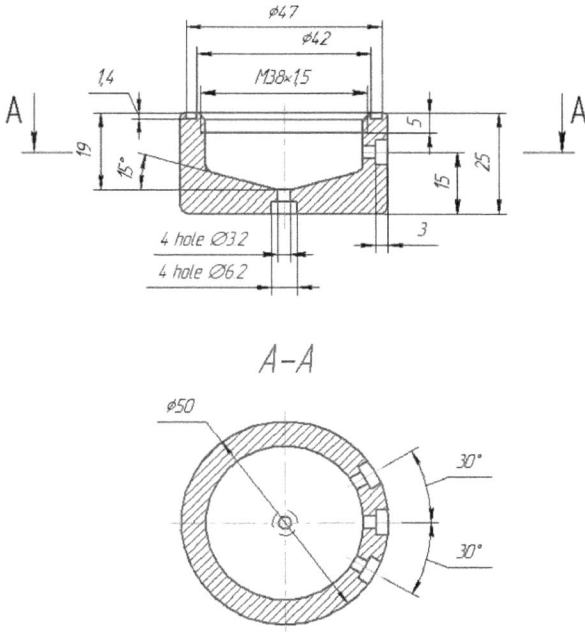

Figure 28. Three-way eddy current mixer for a multicomponent powder mixture [125]

The mixer lid (Figure 28) is designed in such a way that when powders enter the glass, uniform mixing and equalization of the flow rates of the carrier gas and the powder composition occur.

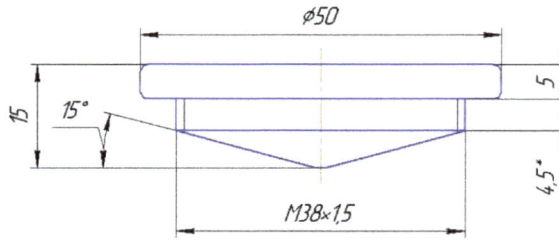

1. *Non-specified rounding radii are assumed to be 2 mm*
2. **Reference dimension*

Figure 29. Mixer cover for a multi-component powder mixture [125]

The presented design of the mixer is interchangeable and can be used depending on the feasibility and purpose of the spraying process.

The functional diagram of the complex for plasma recovery of substandard turbine blades of the CHP is shown in Figure 30.

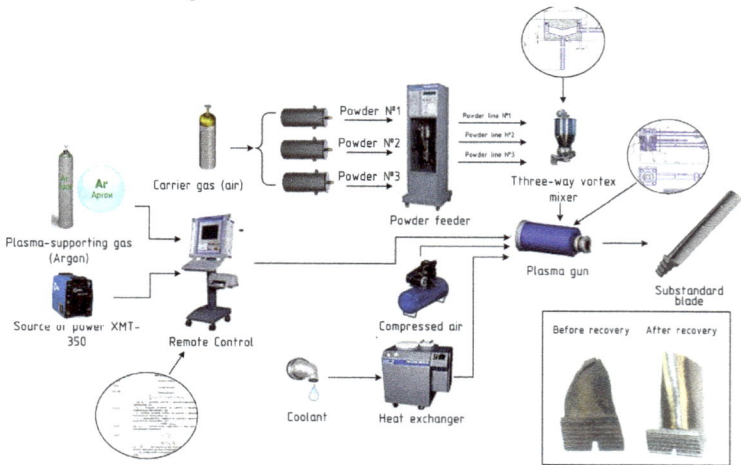

Figure 30. Functional diagram of the connection of the developed mixer and the system as a whole

Modernization of the plasmatron was implemented through the development and implementation of a vortex dispenser, which ensures the efficiency of the use of complex multicomponent powder compositions. This constructive and technological solution allows you to control the physical and mechanical properties of the resulting surface in the process of deposition. The design feature and its attractiveness lies in the presence of a confuser chamber in the design of a three-way vortex dispenser installed at the inlet of flows into the plasmatron nozzle. The turbulence effect of the gas-powder mixture is based on the application of the well-known Bernoulli's laws. This principle of mixture formation increases the adhesive properties of the sprayed material due to the acquired acceleration during droplet impact of particles on the surface to be restored.

Conclusion

1. Research has established that one of the difficult tasks to be solved is not the restoration of the original shape of the product, but the formation of the structure of the base of the blades, steels of the austenitic class. The established cause-and-effect relationships between technological regimes and the mechanical properties of the material structure made it possible to develop an innovative technology for the restoration of substandard turbine blades of thermal power plants. Its distinctive technological feature is the introduction of an implant and surfacing of a stellite plate, followed by thermal cycling.

2. The area of effective values of the parameters of the technological process of restoration, which is characterized by the dependencies of the influence of technological regimes on the quality of restoration, has been studied.

3. In order to control the modes of the technological process of recovery, the plasmatron was modernized by introducing a three-way vortex mixer into the gas-air system. The gas-dynamic calculation of the developed elements of the gas-air path made it possible to design and develop structures. three-way vortex mixer. An analysis of the efficiency of the GWT system of the modernized plasmatron showed an improvement in the gas-dynamic parameters of the TRG flow in comparison with the basic design of the plasmatron by up to 2.8%.

4. The introduction of this design into a complex system of the recovery complex makes it possible to control the physical and mechanical properties of the resulting surface during the deposition process in real time.

Conclusion

After analyzing the standard methods for assessing the reliability of turbine blades, it was found that their essence boils down to theoretical calculations of the acting centrifugal forces, moments of inertia forces and stress state of the root section and blade airfoil in static mode. This approach does not take into account the dynamic moments that occur in the actual operating conditions of the CHP turbine, which reduces their resource life. To improve the accuracy of predicting turbine failures, an integrated technique and algorithm have been developed to determine the vibrational reliability of the blades, taking into account the physical and mechanical properties of their phase structure. This knowledge base increases the reliability of the strength calculation and the interpretation of the obtained measurement results.

The proposed concept made it possible to take into account the actual changes in the mechanical characteristics of the austenitic class material used for the manufacture of rotor blades that have passed a certain period of operation. In turn, fatigue processes in the phase structure of the blade material reflect the negative impact of dynamic loads acting in real operating conditions. The studied processes of degradation of the blade surface made it possible to expand the range of recoverable blades, which were previously rejected.

The substantiation of the negative effect of tensile and compressive stresses in the structure of the material made it possible to develop an innovative technology for restoring both the geometrical parameters and the phase-structural and physical-mechanical properties of CHP turbine blades.

Thus, as a result of the research:

- the mathematical model of dynamic processes in the "steam-blade-turbine" system has been improved, taking into account fatigue stresses and changes in the design geometry due to erosion and corrosion processes in real operating conditions;
- developed a methodology and algorithm for integrated assessment of the quality of restoration of turbine blades of thermal power plants and predicting failures of loaded parts at the pre-fracture stage;
- the dependences of the change in the magnetic field strength of the material of the part on the structural-phase components and parameters of the technological process of plasma recovery of substandard blades are established;
- an innovative technological process was developed for plasma restoration of the base structure and design geometry of substandard CHP turbine blades by the "implantation" method followed by thermal cycling;

- the plasmatron was modernized by introducing a three-way vortex mixer for a multicomponent powder mixture into the gas-air path system, which makes it possible to effectively control the modes of the technological process for the recovery of substandard CHP turbine blades (plasma velocity, composition of the multicomponent powder composition, adhesion).

References

[1] Klubnikin V.S. On achievements in thermal spraying of coatings // Films and Coatings 2001: Proceedings of the 6th International Conference St. Petersburg: Polyplasma, 2001. - P. 15-22

[2] Kostikov V.I., Shesterin Yu.A. Plasma coatings. - M.: Metallurgy, 197. - 159 p.

[3] Lashchenko G.I. Plasma hardening and sputtering. - Kyiv: Ecotechnology, 2003. - 64 p.

[4] Shubenko A.L. Torsional deformations of long blades of steam turbines // Eastern European Journal of Advanced Technologies. - 2013. - 3/8(63). - S. 21-24.

[5] Boyarshinov A.Yu. Improvement of geometry and increase of reliability of fir-tree tail joints of long blades of steam turbines // Problemy mashinostroeniya. - 2014. - V.17, issue 1. - S. 42 - 47.

[6] Bartenev G.M. Strength and fracture of highly elastic materials. - M.-L. : Chemistry, 1964. - 387 p.

[7] Demkin N.B. Surface quality and contact of machine parts / N.B. Demkin, E.V. Ryzhov. - M. : Mashinostroenie, 1981. - 244 p.

[8] GOST 34100.3-2017 Measurement uncertainty. Guidelines for Expressing Measurement Uncertainty - M.: Standartinform, 2017. - Part 3.

[9] Trukhniy A.D. Stationary steam turbines. - M .: Energoatomizdat, 1990. - 640 s

[10] Olkhovsky G.G. Perspective gas-turbine and steam-gas plants for power industry (review) // Teploenergetika, 2013. - №2. - S. 3-11.

[11] Shcheglyaev A.V., Smelnitsky S.G. regulation of steam turbines. - M.: Gosenergoizdat, 1962. -256 p.

[12] Boyko E.A., Bazhenov K.V., Grachev P.A. Thermal power plants (steam turbine power plants TES): a reference guide. - Krasnoyarsk: CPI KSTU, 2006. - 152 p.

[13] Kostyuk A.G., Frolov V.V., Bulkin A.E., Trukhniy A.D. Turbines of thermal and nuclear and electric stations: a textbook for universities. - M.: MPEI Publishing House, 2001. - 488 p.

[14] GOST 23269-78 Stationary steam turbines. Terms and Definitions. - M.: Standartinform, 2005. - 8 p.

[15] Steam turbines. Principle of operation. [Electronic resource]. - Access mode: https://manbw.ru/analitycs/steam-turbines.html. (25.10.2018)

[16] Bauman N.Ya., Novikov V.A. Organization of technological preparation for the

production of steam and gas turbines: a tutorial. - Sverdlovsk: UPI, 1991. - 72 p.

[17] Bushuev M. N. Turbine production technology. - M.; L.: Mashinostroenie, 1966. - 416 p.

[18] Ustyantsev A.M., Nodelman G.I., Novikov V.A. Technology of production of steam and gas turbines: a tutorial. - M.: Mashinostroenie, 1982. - 208 p.

[19] Berezkin V.V., Pisarenko V.S., Michael Yu.S., Benin L.A. Technology of turbine construction. - L.: Mashinostroenie, 1980. - 720 p.

[20] Heat exchangers of power plants: a textbook for universities / ed. Yu.M. Brodov. - Ekaterinburg: Socrates, 2002. - 968 p.

[21] GOST 28969-91. Stationary steam turbines of low power. General specifications. - M.: Publishing house of standards, 1991. - 10 p.

[22] Motorin A.V., Raspopov I.V., Fursov I.D. Steam turbines: textbook: in 2 volumes - Barnaul: AltGTU Publishing House, 2004. - 235 p.

[23] RD 24.260.09 - 87 - RD 24.260.12 - 87 Choice of design, limit deviations of dimensions and roughness parameters of the main structural elements of axial turbomachine blades in the design: a collection of guidelines. - L.: NPO CKTI, 1988. - 36 p.

[24] Shubenko L.A., Shubin D.M., Gerner V.P. Strength of steam turbine elements. - M.: Mashgiz, 1962. - 568 p.

[25] Levin A.V. Working blades and disks of steam turbines. - M .: Gosenergoizdat, 1953. - 624 p.

[26] Levin A.V. Strength and vibration of blades and disks of steam turbines / A.V. Levin, K.M. Borshansky, E.D. Conson. - L .: Mashinostroenie, 1981. - 710 p.

[27] Savinkin V.V., Ratushnaya T.Yu., Kuznetsova V.N., Shakirova M.A. Substantiation of criteria for energy-efficient operation of a CHP turbine, taking into account the unit costs for the restoration of the consequences of a failure // Bulletin of KazNITU. - Almaty, 2019. - No. 5 (135). - pp. 276-285

[28] Kamaraj M. Rafting in single crystal nickel- base super alloys an overview. - Sadhana, 2003. - Vol. 28, № 1-2. - P. 115 - 128. https://doi.org/10.1007/BF02717129

[29] Artamonov V.V., Artamonov V.P. Diagnostics of the causes of the operational destruction of rotor blades of gas turbines // Russian Journal of Nondestructive Testing - 2013. - Vol. 49. - №.9. - P. 538-542. https://doi.org/10.1134/S1061830913090027

[30] Savinkin V.V., Ratushnaya T.Yu., Ivanova O.V., Kenzhetaeva L.D. On the methods of diagnosing hidden defects by FMEA-analysis // Scientific and technical journal "Gosstandart News". - Astana. - 2018. - No. 1 (71). - S. 22-24.

[31] Savinkin V.V., Ratushnaya T.Yu., Ivanova O.V. Statistical analysis of the causes of structural defects in the blades of steam and gas turbines // Proceedings of the MNPK "Science 2017: results, achievements, prospects". - Stavropol: Logos. - 2017. - S. 13-16.

[32] Rozno M.I. How to learn to look ahead? Implementation of FMEA-methodology // Methods of quality management. - 2010. - No. 6. - S. 25-28.

[33] Andreev AV Engineering methods for determining stress concentration in machine parts. - M .: Mashinostroenie, 1976. - 69 p.

[34] Mingazheva A.A., Dautov S.S. Heat-resistant coating with pseudoplastic properties to protect parts from intermetallic alloys 18 // XXXVIII Gagarin Readings. Scientific works of the international youth scientific conference. - M.: MATI, 2012. - S. 47.

[35] Garbul A.F. Influence of mode parameters on the formation of a seam during welding on the weight of butt joints with a plasma arc // Welding production. - 1971. - No. 8. - S. 28-30

[36] Chiang W. C., Pinfold B.E. Operational envelopes for plasma keyhole welded titanium// Welding and Metal Fabrication. -1979. -Vol. 47, № 9. - P.661-673.

[37] Ryzhenkov V.A. State of the problem and ways to improve the wear resistance of power equipment at thermal power plants. Teploenergetika. - 2000. - No. 6. - S. 20-21.

[38] Development and implementation of an energy-efficient technology for the restoration of blades of complex geometry of steam and gas turbines of thermal power plants by highly concentrated sources of plasma energy with an adaptive process control system: a report on research / NKSU named after. M. Kozybaeva: performer: Savinkin V.V., Kolisnichenko S.N., Koptyaev D.A. and others - Petropavtivsk, 2016. - 150 p. - No. GR 0115RK01226.

[39] Patent No. 2251476 Russian Federation. Smyslov A.M., Smyslova M.K., Godovskaya G.V., Isanberdin A.N., Lyudvinitsky S.V., Khafizov R.Kh. Method for restoring steam turbine blades. - No. 2003128016/02; dec. 09/17/2003; publ. May 10, 2005, Bull. No. 13.

[40] Khromchenko F. A., Lapa V. A., Fedina I. V., Dolzhansky P. R. Technology of repair of working blades of steam turbines // Welding production. - 1998. - Part 1,

No. 11. - S. 56 - 70

[41] Gonserovsky F.G. Seventeen years of experience in the operation of steam turbine blades after repair with the use of welding. Teploenergetika - 2000. - No. 4. - S. 39-48.

[42] Gonserovsky FG, Petrenya Yu.K., Silevich VM Efficiency of steam turbine blades repaired by welding // Welding production. - 2000. - No. 1. - S. 10-15

[43] Gonserovsky F. G., Petrenya Yu. K., Silevich V. M. Longevity of steam turbine blades, taking into account repairs in power plants // Electric Stations. - 2000. - No. 3. - S. 45-52

[44] D. N. Korotaev, E. E. Tarasov, K. N Poleschenko, E. N. Eremin, E. V. Ivanova Formation of wear resistant nanostructural topocomposite coatings on metal materials by ionicplasma processing // Journal of Physics: Conference Series. - 2018. - Vol. 1050(1). - R. 012037-1-012037-6. - Doi:10.1088/1742-6596/1050/1/012037/ https://doi.org/10.1088/1742-6596/1050/1/012037

[45] Leshchinsky L.K. Plasma surface hardening. - Kyiv: Technique. - 1990 . - 109 p.

[46] Papshev D.D. Technological methods for improving the reliability and durability of machine parts by surface hardening: a tutorial. - Kuibyshev.: Kuibyshev Polytechnic. in-t im. V.V. Kuibyshev. - 1983. - 81 p.

[47] Savinkin V.V. Increasing the durability of the restored parts of the elements of the hydraulic drive of construction and road machines: dis. ... cand. tech. Sciences. - Omsk, 2009. - 227 p.

[48] GOST 8713-79 Submerged arc welding. Connections are welded. Basic types, structural elements and dimensions. - M.: Publishing house of standards, 1979. -10 p.

[49] GOST 28844-90 Gas-thermal strengthening and restoring coatings. General requirements. - M.: IPK Standards Publishing House, 1990. - 15 p.

[50] Sorokin L.I. Argon-arc surfacing of shrouds of working blades from high-temperature nickel alloys: material of technical information. - M.: Welding production. - 2004. - No. 7. - pp. 36-39

[51] Klimov V.G. Comparison of methods for restoring the geometry of the turbine blade blades from heat-resistant alloys. Bulletin of the Moscow Aviation Institute. - 2016. - V.23, No. 1. - P.86-97.

[52] Kasser D. Laser Powder Fusion Welding. - Electronic resource. Access mode: http://huffman-llc.com/pdf/Articles/LPFW%20Huffman_Kaser.pdf (06/16/2019)

[53] Kathuria Y.P. Some aspects of laser surface cladding in the turbine industry // Surface and Coatings Technology. - 2000. - No. 132. - R.262-269. https://doi.org/10.1016/S0257-8972(00)00735-0

[54] Shepeleva L., Medres B., Kaplan W.D. etal. Laserc lading ofturbine blades // Surfaceand Coatings Technology. - 2000. - Vol.12. - P.45-48. https://doi.org/10.1016/S0257-8972(99)00603-9

[55] Harvin V.S. Scientific and technical developments for the restoration and hardening of machine parts. - M.: Transport, 2002. - S. 27-31.

[56] Tomashets A.K., Savinkin V.V. Justification of the main technological parameters affecting the quality of plasma spraying: materials of the 10th international conference "Physics of the Solid State" // Bulletin of the Karaganda State University. Physics series. - 2008. - No. 2 (50). -- P. 31-38.

[57] Dolzhansky P. R., Dobrokhotov S. E. Improving the operational reliability of the working blades of the last stages of turbines T-250/300-240 // Reliability and safety of energy. - 2008. - No. 1. - S. 56-59

[58] Abhijit Roy A Take Stock of Turbine Blades Failure Phenomenon // Journal of The Institution of Engineers (India): Series C. - 2018. -Vol. 99, Iss. 1. - P. 97 - 103. https://doi.org/10.1007/s40032-017-0375-9

[59] Savinkin V.V., Ratushnaya T.Yu., Abilmazhinova A.A. Studies of the concentration of internal stresses in the turbine blades of a thermal power plant using the metal magnetic memory method // Scientific and technical journal "Metrology". - 2017. - No. 1. - S. 33-42.

[60] 153-34.1-17.462-00. Guidelines on the procedure for assessing the performance of steam turbine rotor blades in the process of manufacture, operation and repair. M.: IPK Standards Publishing House, 2001. - 15 p.

[61] GOST 6996-86 Welded joints. Methods for determining mechanical properties. - M.: Publishing house of standards, 1986. - 10 p.

[62] GOST 7122-81 Welded seams and deposited metal. Sampling methods for determining the chemical composition - M.: Standards Publishing House, 1981. - 9 p.

[63] GOST 9450-76 Measurement of microhardness by indentation of diamond tips - M.: Standards Publishing House, 1976. - 8 p.

[64] Dubov A.A. Diagnostics of turbine equipment using metal magnetic memory. - M., 2009. - S. 148.

[65] Chen Ni, Lin Hua, Xiaokai Wang, Zhou Wang, Xunpeng Qin, Zhou Fang Coupling method of magnetic memory and eddy current nondestructive testing for retired crankshafts // Journal of Mechanical Science and Technology. - 2016. - Vol. 30, Iss. 7. - P. 3097 - 3104 https://doi.org/10.1007/s12206-016-0618-3

[66] Shangkun R., Xianzhi R., Zhenxia D., Yuewen F. Studies on influences of initial magnetization state on metal magnetic memory signal // NDT & E International. - 2019. - Vol. 103. - P. 77-83. https://doi.org/10.1016/j.ndteint.2019.02.002

[67] Hailong Ch., Changlong W., Xianzhang Z. Research on methods of defect classification based on metal magnetic memory // NDT & E International. -2017. - Vol. 92. - P. 82-87 https://doi.org/10.1016/j.ndteint.2017.08.002

[68] Dubov A.A. Metrological aspects in the method of metal magnetic memory // World of Measurements. - M., 2018. - No. 4. - S. 41-43

[69] ArteagaC., RodríguezJ.A., ClementeC.M., SeguraJ.A., UrquizaG., HamzaouiY.El. Estimation of use fullifein turbines blades with cracks in corrosive environment // Engineering Failure Analysis. - 2013. - Vol. 35. - P. 576-589 https://doi.org/10.1016/j.engfailanal.2013.05.013

[70] KováříkaO., HaušildaP., SieglaJ., MatějíčekbJ., DavydovcV. Fatigue Lifeof Layered Metallicand Ceramic Plasma Sprayed Coatings // Procedia Materials Science. - 2014. - Vol. 3. - P. 586-591 https://doi.org/10.1016/j.mspro.2014.06.097

[71] Shubenko L.A. Strength of steam turbines. - M.: Mashinostroenie, 1973. - 456 p.

[72] Shorr B.F. Calculation of the strength of naturally swirling blades. - M.: MAP, 1954. - No. 256. - 18 s.

[73] Vorobyov Yu.S, Shorr B.F. Theory of twisted rods - Kyiv: Naukova Dumka, 1983. - 188 p.

[74] Subbotin V.G., Levchenko E.V., Shvetsov V.L., Shubenko A.L., Tarelin A.A., Subbotovich V.P. Creation of steam turbines of a new generation. - Kharkov: Folio, 2009. - 256 p.

[75] Shorr B.F. Fundamentals of the theory of twisted blades with an indirect axis // Strength and dynamics of aircraft engines. - 1966. - Issue. 3. - S. 188-213.

[76] Boyarshinov A.Yu. Improving the design of the working blades of the last stages of steam turbines in order to improve their performance: diss. ... Candidate of Technical Sciences. - Kharkiv. - 2016. - 137 p.

[77] Velikanova N.P., Zakiev F.K. Comparative analysis of the strength reliability of turbine blades for aircraft gas turbine engines with a long service life. Vestnik

dvigatelestroyenija. - 2006. - No. 3. - S. 80-83.

[78] Larson F.R., Miller J. Atime-temperature relationship for rupture and creepstresses //
Trans. ASME. - 1952. - Vol. 74. - P. 765-775. https://doi.org/10.1115/1.4015909

[79] Demyanushko I.V., Velikanova N.P., Kornoukhov A.A. Forecasting the durability of
rotor parts taking into account real loading // Aerospace Technique and
Technology. - 2001. - No. 23. - pp. 119-120

[80] Manson S.S. Temperature stresses and low-cycle fatigue. - M.: Mashinostroenie,
1974. - 450 p.

[81] Korolev A.N., Velikanova N.P., Zakiev F.K. Influence of operational factors on the
durability of turbine disks of aircraft gas turbine engines // Aerospace Technique
and Technology. - 2001. - No. 23. - pp. 116-118

[82] Velikanova N.P., Kiselev A.S. Analysis of the influence of operating time on the
durability of the heat-resistant alloy of turbine rotor blades // Bulletin of KSTU im.
A.N. Tupolev. - 2001. - No. 1. - pp. 23-26

[83] Aleksandrov A.V., Laschenikov B.Ya., Shaposhnikov H.H. Structural mechanics.
Thin-walled spatial systems. - M .: Stroyizdat, 1983. - 488 p.

[84] Bogdanovich A.E. Nonlinear Problems of the Dynamics of Cylindrical Composite
Shells. - Riga: Zinatne, 1987. - 256 p.

[85] Bublik B.I. Numerical solution of dynamic problems of the theory of plates and
shells. - Kyiv: Nauk. Dumka, 1976. - 222 p.

[86] Weinberg D.V., Gorodetsky A.S., Kirichevsky V.V. Sakharov A.S. Finite element
method in the mechanics of deformable bodies // Applied Mechanics. - 1972. - T.
VIII, no. 8. - S. 3-28. https://doi.org/10.1007/BF00886062

[87] Birger I.A., Schorr B.F., Shneiderovich R.M. Calculation of the strength of machine
parts (reference guide). - M.: Mashinostroenie, 1966. - 616 p.

[88] Golovanov A.I., Berezhnoy D.V. Finite element method in mechanics of deformable
solids. - Kazan: DAS, 2001. - 86 p.

[89] Zhigalko Yu.P., Shigabutdinov A.F. To the question of the stability of elements such
as rods and shells under longitudinal dynamic loading // VIII All-Russian Congress
on Theoretical and Applied Mechanics. - Perm. - 2001. - P.259.

[90] Levin A.V., Borishansky K.N., Konson E.D. Working blades and disks of steam
turbines // Scientific and technical statements of the St. Petersburg Polytechnic
University. 2013. - No. 2 (171). - pp. 52-60

[91] Levin A.V., Shur S.S. Torsional vibrations of working blades connected in a package

// Energomashinostroyeniye. - 1961. - No. 8. - S. 1-4.

[92] Leikin A.S. Tension and Endurance of Details of Complex Configuration // Mashinostroenie. - 1968. - 71 p.

[93] Moiseev N.N. Asymptotic methods of nonlinear mechanics. M.: Nauka, 1981. 325 p.

[94] Pavlov V.P. Modeling on a computer of loads in the elements of hydraulic mechanisms of an arbitrary structure // Vehicles of Siberia: materials of an interuniversity scientific and practical conference with international participation. - Krasnoyarsk: Publishing House of KSTU, 1995. - S. 326 - 330

[95] Golovanov A.I., Shigabutdinov A.F. To the calculation of free vibrations of thin-walled structures // Proceedings of the 13th interuniversity. conf. - Samara, 2003. - Part 1. - S. 28-32.

[96] Gazimov M.M., Golovanov A.I., Shigabutdinov A.F. Calculation of thin-walled structures for free vibrations by the finite element method // Sat. mater. XV All-Russian. interuniversity scientific and technical conf. - Kazan, 2003. - Part 1. - S. 328-329.

[97] Ivanov V.P. On the question of the causes of the spread of resonant stresses in elastic bodies with structural cyclic symmetry // Proceedings of KAI. - 1969. - Issue. XXSH. - S. ShZ-19.

[98] Uriev E.V. Vibration reliability of steam turbines and methods for its improvement: Ph.D. ... Doctor of Technical Sciences: 05.04.12.- M., 1997.- 40 p.

[99] Zablotsky I.E., Korostelev Yu.A., Shipov R.A. Non-contact measurements of oscillations of turbomachine blades. - M.: Mashinostroenie, 1977.-159 p.

[100] Kiselev M.I., Morozov A.N., Nazolin A.L. et al. Transitional processes during the start-up of a turbogenerator // Engineering-physical problems of aviation and space technology: abstracts. report 2nd intl. sci.-tech. conf. - Egorievsk, 1997. - S. 60-61.

[101] Kairov A.S., Morgun S.A. Investigation of oscillations of a disk with a crown of rotor blades of turbomachines as a cyclically symmetric system in the field of centrifugal forces. Vestnik dvigatelestroeniya. General questions of engine building. - 2013. - No. 1. - pp. 34-37

[102] Savinkin, V.V., Kuznetsova, V.N., Ratushnaya, T.Yu., Kiselev, L.A. Method of integrated assessment of fatigue stresses in the structure of the restored blades of CHP and HPS // Bulletin of the Tomsk Polytechnic University, Geo Assets Engineering. - 2019. - Vol. 330. - № 8. - P. 65-77. DOI: 10.18799/24131830/2019/8/2213 https://doi.org/10.18799/24131830/2019/8/2213

[103] Savinkin V.V., Ratushnaya T.Yu. Mathematical description of the main factors, increasing the gas-dynamic load of the CHP turbine // Bulletin of the NKGU im. M. Kozybaeva. Series technical. - 2016. - No. 4 (33). - S. 114-118.

[104] Savinkin V.V., Kuznetsova V.N., Ratushnaya T.Yu., Kiselev L.A. Investigation of fatigue stresses in the phase structure of the blade airfoil and assessment of the resource reliability of the turbine. Scientific journal Vestnik Mashinostroeniya. - 2019. - No. 6. - S. 34-40.

[105] Redozubov A. Yu., Krivonogova A. S. Quality control of turbine blades in the production process // Molodoy ucheny. - 2016. - No. 12.3. - S. 68-71. - URL https://moluch.ru/archive/116/31855/ (date of access: 11/25/2018).

[106] GOST 34497 - 2018 Steam turbine blades. Basic requirements for replacement.: M: Standartinform, 2019. - 28 p.

[107] Petrov, Yu.V. On the temperature dependence of the threshold rate of erosion destruction / Petrov Yu.V., Smirnov V.I. // Reports of the Academy of Sciences. - 2007. - T. 416, No. 6. - P. 766-768.

[108] Savinkin V.V., Kuznetsova V.N., Ratushnaya T.Yu., Kiselev L.A. (2019) Method of integrated assessment of fatigue stresses in the structure of the restored blades of CHP and HPS. Bulletin of the Tomsk Polytechnic University, Geo Assets Engineering 330(8):65-77. DOI: 10.18799/24131830/2019/8/2213 https://doi.org/10.18799/24131830/2019/8/2213

[109] Skotnikova M.A. The use of titanium alloys as a material for steam turbine blades. Voprosy materialovedeniya. - 2007. - No. 3 (51). - S. 61-70.

[110] Redozubov A. Yu., Krivonogova A. S. Quality control of turbine blades in the production process // Molodoy ucheny. - 2016. - No. 12.3. - S. 68-71. - URL https://moluch.ru/archive/116/31855/ (date of access: 11/25/2018).

[111] GOST 34497 - 2018 Steam turbine blades. Basic requirements for replacement.: M: Standartinform, 2019. - 28 p.

[112] Ratushnaya T.Yu., Savinkin V.V, Tomashets A. K., Tyukanko V.Yu Substantiation of criteria for evaluation of substandard turbine blades of CHP in the process of rejection // Scientific and technical journal "Metrology". - 2019. - No. 4 (71). - p. 47-50.

[113] Savinkin V.V, Vizureanu P., Sandu A.V., Ratushnaya T.Yu., Ivanischev A.A, Surleva A. Improvement of the Turbine Blade Surface Phase Structure Recovered by Plasma Spraying // Coatings. - 2020. - Vol. 10, Iss.1. - P. 62 https://doi.org/10.3390/coatings10010062

[114] Ratushnaya T.Ju., Savinkin V.V Vizureanu P., Sandu A.V.,Ivanischev A.A. Improvement of the Turbine Blade Surface Phase Structure Recovered by Plasma Spraying // Improvement of the Turbine Blade Surface Phase Structure Recovered by Plasma Spraying // Coatings. - 2020. - V 10. - Iss.1. - PP. 62 (Q2). doi:10.3390/coatings10010062; www.mdpi.com/journal/coatings. https://www.mdpi.com/2079-6412/10/1/62/htm, ISSN: 20796412, DOI: 10.3390/coatings10010062 https://doi.org/10.3390/coatings10010062

[115] GOST 25.507-85. Metals. Test method for high-cycle and low-cycle fatigue. - M: Standartinform, 1986. - 10 p.

[116] GOST 24346 - 80 Vibration. Terms and Definitions. - M: Standartinform, 1981. - 8 p.

[117] Savinkin V.V., Ratushnaya T. Ju., Kiselev L.A. Substantiation of efficiency of plasma recovery of physical and mechanical properties of turbine blades of CHP // Bulletin of the SemGU im. Shakarim. - Semey, 2019. - No. 1(85) - S. 92-95

[118] Anakhov S.V. Development of scientific principles and methods for designing plasmatrones to improve the efficiency and safety of electroplasma technologies: diss. ... doc. tech. Sciences. - Ekaterinburg. -2019. - 291 p.

[119] Klimenko A.A., Lyapin G.K. Designs of electric arc plasmatrones. - M .: Izdvo MSTU im. Bauman, 2010. - 56 p.

[120] Anshakov A.S., Dandarov G.-N. B., Efremov V.P. et al. Electric arc plasmatrones: advertising brochure / ed. M.F. Zhukov. -Novosibirsk: Institute of Thermal Physics, 1980. - 84 p.

[121] Rutberg F.G., Safronov A.A., Popov S.D. et al. Multiphase electric arc plasmatrons of alternating current for plasma technologies // TVT. - 2006. - T. 44, No. 2 - S. 205-211. https://doi.org/10.1007/s10740-006-0024-0

[122] Cherednichenko V.S., Anshakov A.S., Kuzmin M.G. Plasma electrotechnological installations. - Novosibirsk: Publishing house of NSTU, 2011. - 602 p.

[123] Anakhov S. Gas and RRR distribution in high purity Niobium EB welded in ultra-high vacuum / S. Anakhov, X. Singer, W. Singer, H. Wen. - New-York: Proceedings ISOHIM-2005, -2006. - Vol.837 - C. 71-85 https://doi.org/10.1063/1.2213061

[124] Pykin Yu.A. Efficiency and energy saving - criteria for choosing electroplasma technologies / Pykin Yu.A., Anakhov S.V. - Ural Federal District: Construction. ZhKK. - 2010. - No. 1. - S. 22-23.

[125] Ratushnaya T.Yu., Savinkin V.V., Ivanova O.V., Shakirova M.A., Patent for utility

model No. 6809 "Three-way vortex mixer" Republic of Kazakhstan / Three-way vortex mixer / No. 2021/1113.2 application. 08.12.2021; publ. 01/14/2022

[126] Savinkin V.V., Ratushnaya T.Yu., Kuznetsova V.N. Study of the mechanical parameters of the turbine blades of thermal power plants, restored by a plasma source of energy // Bulletin of KazNITU. - Almaty, 2018. - No. 6 (130). - S. 337-345.

[127] V. V. Savinkin, T. Yu. Ratushnaya, A. A. Ivanischev, A. R. Surleva, O. V. Ivanova, S. N. Kolisnichenko Study on the Optimal Phase Structure of Recovered Steam Turbine Blades Using Different Technological Spray Modes for Deposition of Al2O3 // The 5th International Conference on Green Design and Manufacture 2019 IConGDM 2019. - Bandung, Indonesia 2019. - P. 64-68. - 2129, 020022 (2019); индексируется Web of Science, DOI:10.1063/1.5118030 https://doi.org/10.1063/1.5118030

[128] Savinkin V.V., Ratushnaya T.Yu., Ivanishchev A.A., Belyi A.V., Kovalchuk E.N. Modern methods and technologies for the creation and processing of materials" based on the results of a scientific internship Study of the properties and structure of coatings obtained by plasma spraying using Al2O3 // Modern methods and technologies for the creation and processing of materials - T2. Technologies and equipment for mechanical and physical-technical processing. - Minsk, 2018. - S. 229-234.

[129] Ratushnaya T.Yu., Savinkin V.V., Zykova N.V. Determination of the list of quality indicators for blades of steam and gas turbines of CHPPs, restored by a highly concentrated source of plasma energy // 2nd International modern scientific research congress, December 23-25, 2021/Istambul, Turkey, p. 528